A Computer Perspective

*A sequence of 20th century ideas,
events, and artifacts from the history
of the information machine*

"A Computer Perspective." Exhibition designed by the office of Charles and Ray Eames, IBM Exhibit Center, New York City

A Computer Perspective

By the office of
Charles & Ray Eames

Edited by Glen Fleck

Produced by Robert Staples

Introduction by I. Bernard Cohen

Harvard University Press, Cambridge, Massachusetts, 1973

Book Design by Paul Bruhwiler, Inc.
for the office of Charles and Ray Eames

A Computer Perspective is an illustrated essay on the origins and first lines of development of the computer. The complex network of creative forces and social pressures that have produced the computer is personified here in the creators of instruments of computation, and their machines or tables; the inventors of mathematical or logical concepts and their applications; and the fabricators of practical devices to serve the immediate needs of government, commerce, engineering, and science.

The book is based on an exhibition conceived and assembled for International Business Machines Corporation. Like the exhibition, it is not a history in the narrow sense of a chronology of concepts and devices. Yet these pages actually display more true history (in relation to the computer) than many more conventional presentations of the development of science and technology.

A Computer Perspective reveals the ways in which dreamers and theorists encountered engineers and practical inventors. The fruits of their joint efforts are seen to have enriched the modes of solving problems of business and industry, science and technology, education and psychology, and even the functioning of government.

The *Perspective* has been broadly conceived so as to display aspects of the intellectual and socio-economic environments of the sixty years that have led to the creation of the modern computer. We are made aware of the historical periods or backgrounds by a continuing display of pictorial materials. A novel by E. M. Forster or by Aldous Huxley (reminding us of the general atmosphere of concern about the new world of machines, robots, and automation) or a campaign button, a page of popular sheet music, or a hubcap from an ancient Ford automobile, may give us at once the time sequence, decade by decade, in which the steps toward the computer were made.

Here is a revelation of the varied human dimensions of the computer's origins, showing the many aspects and influences of abstract thought, mechanical invention, and hardheaded business practice that have ended up in the computer. We are reminded that the computer is the product of men's minds and hands, and that the manifest complexities of its influence upon our lives reflect the incredible variety and complexity of sources from which it has sprung.

Even the growth of psychology finds a proper place here. In its early stages, and even now, the computer has been conceived as a mechanical brain or a mechanized intelligence. Furthermore, the converting of the concept of intelligence into numbers has continually provided a vast area for statistical analysis. Similarly, the discovery that the human body is a self-regulating device cannot be ignored in tracing the history of feedback mechanisms. Indeed, one of the most significant aspects of this essay may be the revelation of the complex set of interactions between the concepts and practices of psychology and physiology and the art and design of the computer—the science of man and the products of human ingenuity providing models to one another for advances in both knowledge and design.

This book identifies and traces the history of three major classes of innovation (whether of concept and theory, or practice and device) that have interacted and fused to create the computer in the 1940s. Designated "logical automata," "statistical machines," and "calculators," these key words (with some variations) occur throughout the text, enabling the reader to trace the historical evolution of the concept or practice of automated or self-controlling and self-regulating devices; of machines that sort or process information; and of instruments of calculation.

These three thematic lines of development, coalescing in the production of the computer as we know it, bring a unity to such otherwise disparate entries as Baum's *The Wizard of Oz,* the chess-playing machine designed by Torres y Quevedo, business machines for accounting or billing, gyroscopes, books on logic, and trigonometric tables. In some instances there may not appear to be a close fit between these captions and the concept or instrument under discussion: the relationship may be one that becomes apparent only much later and in retrospect. Galton was not actually associated with "statistical machines" as such, but all aspects of modern statistics (and thus statistical machines) have been so influenced by him that we have linked his name with a category that fits only loosely. The same is true of Taylorization in relation to automation, of Binet and the development of the I.Q. in relation to statistical analysis of human capabilities, or of operations research in relation to automata or programmed sequences of activity that may optimize the chances of a desired outcome.

Like the exhibition on which it is based, the book begins in the final decade of the last century and concludes at the end of the fifth decade of the twentieth century. The beginning date has a cogent logic of its own, because the United States Census of 1890 was the stimulus for the invention and introduction of Herman Hollerith's electro-mechanical sorting and data-processing devices, using punch cards: the first major statistical machines to be built and put into large-scale use. With this invention the information-handling revolution was launched.

The terminal date of 1950 marks the time when the computer revolution had been fully inaugurated with the first generation of modern computers, based on the fundamental concept of the stored program. The concept of a self-regulating machine was far from new in the 1940s; in the 1830s Charles Babbage had already begun the design of an Analytical Engine that would perform mechanically at least four major functions of all computing by human beings: carrying out arithmetic operations, having a "memory," making a choice of computing sequence (which the engine would do automatically), and being capable of numerical input and output. But the stored program, as conceived by John von Neumann in the 1940s, was far more profound an innovation than mere self-regulation or the automation of choices at various stages of computation.

The von Neumann concept gave the computer the capability of treating its regulatory instructions in the same fashion as the numerical data of the input, so that the computer could alter its program and make such logical decisions as might be required from moment to moment in the course of its operation. Almost at once an unanticipated flexibility of operation and purpose was disclosed, with the result that the computer became a dominating element in almost every activity of our society—government, industry, commerce, science, education, and social science and planning.

No longer a specialized instrument designed for a specific task or group of tasks in pure or applied science and engineering (often associated with military needs), the computer emerged around 1950 as a general tool whose effects on all aspects of our daily lives future historians must make precise and evaluate. But for now, the task of making clear the historical forces that produced the modern computer has been elegantly accomplished in *A Computer Perspective*.

A Note on the Exhibition

The exhibition upon which this book is based was designed to illuminate the creative forces and social impact of history being made in our own times. Designed by the Office of Charles and Ray Eames, the exhibition demanded new or refined techniques of communication, based upon such radical innovations as an in-depth correlation of printed explanation, artifact, document, photograph, and quoted text.

The complex origins and influences of the computer were displayed in a "history wall," made up of six 8-foot-high panels, one for each decade from 1890 to 1949. For each decade, there were mounted in a three-dimensional grid a set of documents, photographs, and actual objects, including scientific instruments, small and large calculators, business machines, data processing devices, and the input or output of such machines.

A visitor walking along the wall would find certain objects at the rear of each case disappearing from view while others would emerge, giving to the whole exhibition the kinetic effect of the actual passage of time. The large number of objects, photographs, and documents enabled the viewer to appreciate the variety and multiplicity of closely interrelated causes and effects which have marked the development of the computer; a series of guides and a system of color coding showed the main lines of development which only patient study and research may unravel from historical events.

The techniques of the exhibition enabled each viewer to simulate on a small scale the life of science and invention that have produced the electronic digital computer.

I. B. C.

8

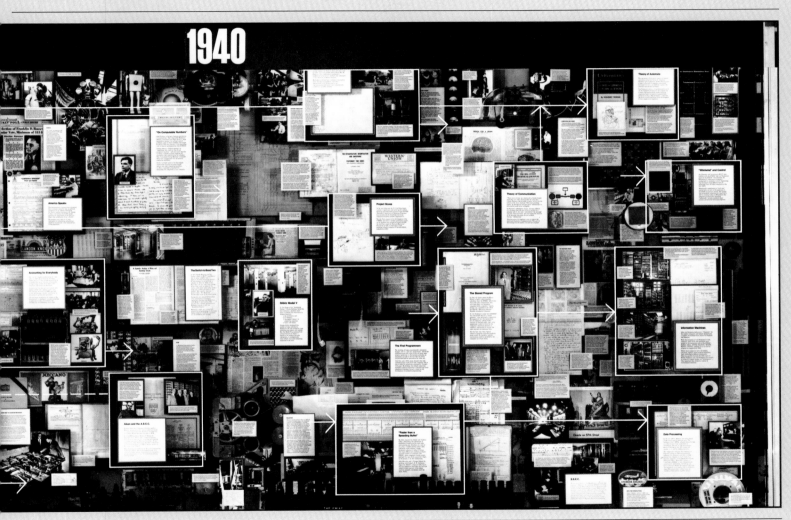

Babbage's Analytical Engine

There had been a single, clear glimpse of the computer idea when Charles Babbage designed an "Analytical Engine" about a hundred years before the first modern computer appeared.

His machine was never built, but it combined the essential parts of the computer idea. His machine would calculate, it would process statistics, and it would automatically guide its own actions based on the answers it was producing.

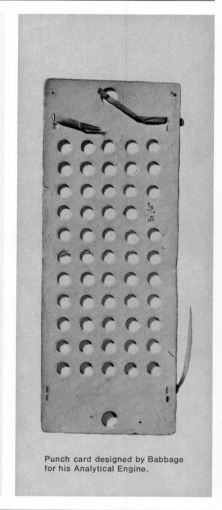

Punch card designed by Babbage for his Analytical Engine.

Perhaps it is because Charles Babbage's work belongs more to our time than to his own that it remains of special interest. While he is recognized today largely for his work on calculating machines, he made other enduring contributions as well. His sustained effort to secure government subsidies helped set an important precedent, and his book, *On the Economy of Machinery and Manufacturers,* laid the groundwork for what we know today as operations research.

As a child in Somerset, Babbage's health was poor, and his education at home encouraged such interest and freedom in the field of mathematics that later, at Trinity College, his tutors were a disappointment. As undergraduates, Babbage, John Herschel, and George Peacock founded The Analytical Society, promising each other to "do their best to leave the world wiser than they found it."

Drawn to intellectual societies, Babbage was involved in the founding of the Royal Astronomical Society. Among his friends were Thomas Carlyle, Charles Darwin, Charles Dickens, Pierre Simon de Laplace, Sir Marc Isembard Brunel, Sir George Everest, and the Countess of Lovelace (Lord Byron's daughter), who, through her understanding of mathematics, machines, and the Babbage theories, has been able to pass on some of the most intelligible accounts of Babbage's work.

Babbage's ideas were so advanced and his standards so high that most efforts to realize his plans during his own lifetime were bound to be unsuccessful. Because of this, many view his life as a series of disappointments, missing the magnitude of his productivity and the breadth of his vision.

Calculating Machines
Statistical Machines
Logical Machines

It was the bringing together of these three lines of machine development,
in the middle of the twentieth century, that made possible an entirely new kind
of machine—the electronic digital computer.

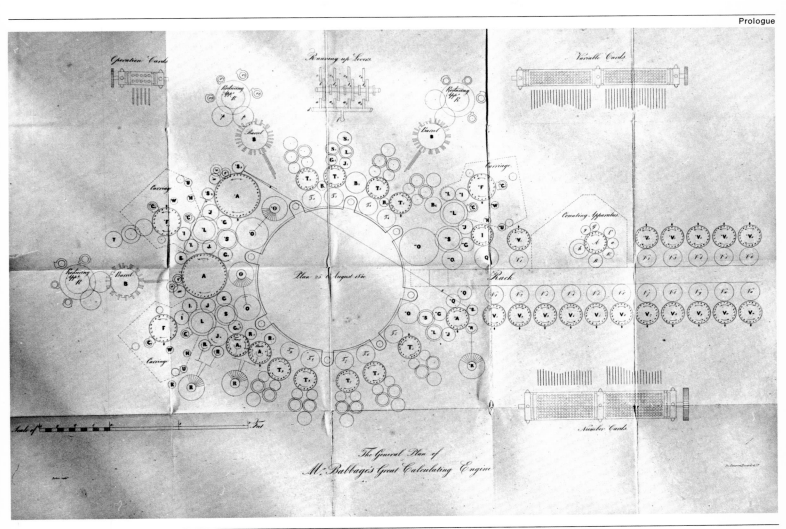

The General Plan of
Mr. Babbage's Great Calculating Engine

While working on a ''special-purpose'' calculator in 1833—his Difference Engine—Babbage started the design for a massive device which, if built, would have been the first general purpose machine: the Analytical Engine. Capable of doing any mathematical operation, it would follow the instructions programmed into it by its operators, and even go on to make decisions about which instructions to follow next, based on the results of its own work.

The Analytical Engine's design contained a great many features we now associate with the modern computer. Both information and instructions were entered on punch cards, and stored in a memory (which Babbage called the ''store''). Following instructions on the operation cards, a processing unit (the ''mill'') performed operations on the information, and returned the results to the ''store.'' The final results were to be printed out, or automatically set in type.

The Analytical Engine was conceived to be on a grand scale. Powered by steam, it would store a thousand fifty-digit numbers in its memory. When the machine needed additional values for the calculation in progress, it was to signal its operators by ringing a bell.

Calculating Machines

Machines used in performing mathematical operations—for commerce, tablemaking, and scientific analysis.

Adding machines and arithmometers were fairly common toward the close of the nineteenth century. The procedures for multiplication and division were slow and indirect, however, so lengthy computations were generally done with logarithmic tables.

Analog devices were calculating tides in Britain and the United States, and a few complicated "difference engines" had seen limited use.

Prologue

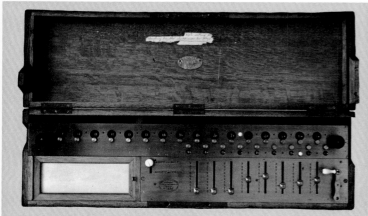

Thomas arithmometer, c. 1850

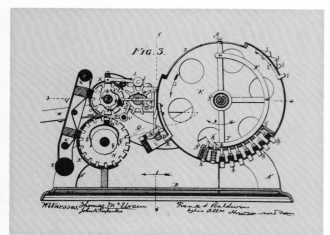

Baldwin calculator, 1875

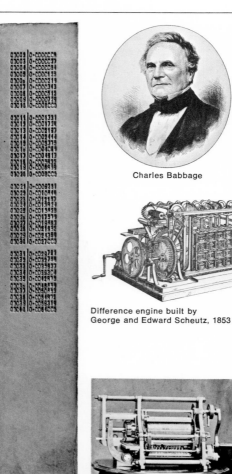

Charles Babbage

Babbage's difference engine

Difference engine built by
George and Edward Scheutz, 1853

George Scheutz

Type mold made by
the Scheutz engine

Wiberg difference engine, 1859

Calculators

Calculating machines were in limited use by the 1890s. Those of Thomas de Colmar were the most popular—several thousand had been built starting about 1830—with Odhner's pinwheel principle running a close second.

Difference Engines

These machines, which accumulated differences to produce tables for navigation, insurance, and astronomy, began with Charles Babbage's design in 1823. Other machines based on his concept followed. Perhaps five had been built by the turn of the century.

John Couch Adams

Urbain J. J. Leverrier

Kelvin's tide predictor, 1873

Sil William Thomson, Lord Kelvin

15

Découvertes nouvelles, caricatures par Cham.

Cartoon showing the discovery of Neptune
by Adams and Leverrier, from L'Illustration, 1846

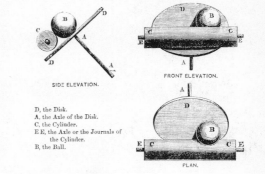
James Thomson's ball and disk integrator, 1876

Methods of Computation

In the late eighteenth century, new computational techniques were devised to solve the large-scale calculations required in astronomy and ballistics. Astronomer John Couch Adams devised a method of solving differential equations in 1883. This computational technique was used by Francis Bashforth to improve the accuracy of ballistic tables.

Analog Computing

A ball and disk integrator was the vital invention needed to build the first automatic analog computing machines. Lord Kelvin used this integrator—devised for a planimeter in the 1860s by his brother, James Thomson—on two new kinds of analog computers: a harmonic analyzer and a tide predictor. He later specified a more general machine—a differential analyzer.

Statistical Machines

Machines used to find relationships within information by sorting and tabulating masses of data.

A growing interest in social investigation had, by the 1890s, created a passion for the statistical method. Techniques that would be useful in organizing the large amounts of information being accumulated by government, business, and the sciences were developed.

The first data processing machine had just been built—the statistical tabulator of Herman Hollerith.

Prologue

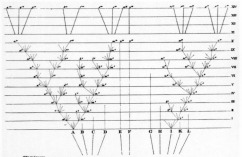

Chart from *On the Origin of Species by Means of Natural Selection*, 1859

Charles Darwin

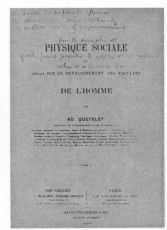

Adolphe Quetelet

Title page of Quetelet's book presented to Florence Nightingale

"My mind seems to have become a kind of machine for grinding general laws out of large collections of facts."
— Charles Darwin

Sir Francis Galton

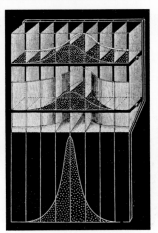

Galton's pin-machine illustrating the effect of natural selection, 1877

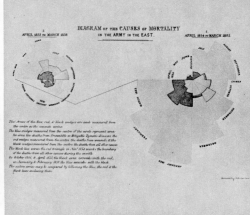

"Coxcombs"—Florence Nightingale's system for displaying statistics

Florence Nightingale

Statistical Science

In science the statistical point of view was still relatively new. The most notable example—Darwin's Theory of Natural Selection—influenced, among others, Sir Francis Galton.

Social Statistics

A general interest in statistics as a force for social reform was inspired by the work of Belgian statistician Adolphe Quetelet. The "Lady of the Lamp," Florence Nightingale, showed such enthusiasm for the use of statistics (she wanted to test the effectiveness of social legislation) that she was termed the "Passionate Statistician."

General view of a company office, 1889

Map from 1870 census showing birth rate

Francis A. Walker

A checking room

President James A. Garfield

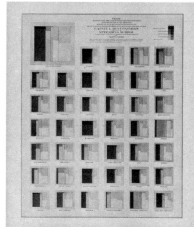
Chart from the 1870 census showing
distribution of occupations by state

Bookkeeping and Statistics

Handling the information needed to run a business—sales figures, inventory, billing, and profit and loss statements—was done by clerks, by hand. But by 1890 the growing size of businesses was straining the capacity of manual systems.

The Census

The U.S. Constitution calls for a census every ten years to establish congressional representation. But Francis Walker, directing the 1880 census, placed significant emphasis on collecting information beyond that required by the Constitution, including for the first time a comprehensive census of manufactures.

Logical Automata

Machines that use information about their past performance to determine their next actions.

By 1890 logical automata, a class of machines that would combine logical decisionmaking with automatic control, had not yet appeared.

There were self-regulating automata, using such feedback mechanisms as the ball governor, and logic machines, such as Jevons' Logical Piano, but the two ideas had not been combined in a single type of machine.

Prologue

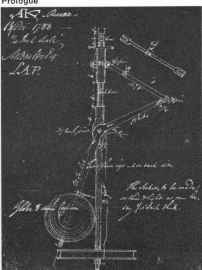

Ball governor

James Watt

The elaborate gears and compartments in von Kempelen's chess player cleverly concealed the man hidden inside.

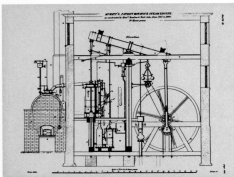

Watt's steam engine with ball governor, 1787

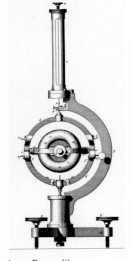

Leon Foucault's gyroscope, 1852

"Are we not ourselves creating our successors . . . daily giving them greater skill and supplying more and more of that self-regulating, self-acting power which will be better than any intellect?"
Samuel Butler, *Erewhon*, 1872

Samuel Butler

18

Feedback and Control

Controlling a machine with feedback—diverting part of the machine's work to direct the work it is about to do—became a familiar idea with the ball governor in James Watt's steam engine, and in 1830 Andrew Ure invented, in principle, the thermostat. Shortly after, Charles Babbage designed a method for his Analytical Engine, in which the result of one set of calculations determines which calculations the machine should do next.

Attitudes Toward Machines

The popularity of mechanical automata, whether simple cam-operated figures or elaborate illusions such as the Wolfgang von Kempelen Chess Player, and the rapid growth of industrial machinery led to some early visions of an automated world. Samuel Butler began his futuristic novel *Erewhon* with an essay, "Darwin Among the Machines."

J. M. Jacquard

> "the Analytical Engine *weaves algebraical patterns* just as the Jacquard loom weaves flowers and leaves."
> — Ada Augusta, Countess of Lovelace (writing about Babbage's proposed machine)

Jacquard Loom

Jacquard loom card

George Boole

DEMONSTRATOR,
INVENTED BY
CHARLES EARL STANHOPE.

The right-hand edge of the gray points out, on this upper scale, the extent of the gray, in the logic of certainty.

The lower edge of the gray points out, on this side scale, the extent of the gray, in the logic of probability.

The area of the square opening, within the black frame, represents the holon, in all cases.

The right-hand side of the square opening points out, on this lower scale, the extent of the red, in all cases.

The right-hand edge of the gray points out, on the same lower scale, the extent of the consequence, (or dark red,) if any, in the logic of certainty.

Rule for the Logic of Certainty.
To the gray, add the red, and deduct the holon: the remainder, (or dark red,) if any, will be the extent of the consequence.

Rule for the Logic of Probability.
The proportion, between the area of the dark red and the area of the holon, is the probability which results from the gray and the red.

PRINTED BY EARL STANHOPE, CHEVENING, KENT.

The first logic machine, 1777

Jevons' logic machine

William Stanley Jevons

Programmed Machines

A line of punched paper cards automatically controls the patterns woven by the Jacquard loom, invented about 1800; but even earlier some musical instruments had been programmed to perform, controlled by rolls of punched paper.

Logic Machines

After George Boole published his method for solving problems in logic in 1854, a number of logic machines based on it were built. The first to solve a problem faster than could be done by hand was Jevons' Logical Piano in 1869.

21

The 1890 Census

The Eleventh U. S. Census posed a crisis in data processing. Figures from the 1880 census were still being interpreted in 1887. At that rate, especially in view of population increase, the 1890 figures would be obsolete before they could be completely analyzed.

But population increase was not the only problem. Although the census had begun as a mere counting of heads (as required by the Constitution), it had through the years become a complex inventory of such population characteristics as immigration, health, racial composition, literacy, and employment.

1890   Statistical Machines

22

Immigrants arrived at Ellis Island in ever-increasing numbers. Cities were growing and there were mass movements westward. To keep up with these rapid changes the government—and private concerns—needed to get more complete information in less time than ever.

Herman Hollerith and His Electric Tabulating Machines

Unless a fast, accurate form of mechanical tabulation were found, the scope of the 1890 census would have to be narrowed. The Census Office held a competition to select an efficient census-taking system.

Herman Hollerith won the contest with his electric tabulating system, which did the job in less than half the time required by either of the rival systems.

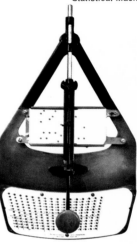

William C. Hunt

Charles F. Pidgin

"I will have in future years the satisfaction of being the first 'statistical engineer.'"
Herman Hollerith

REPORT OF A COMMISSION

HONORABLE SUPERINTENDENT OF CENSUS

DIFFERENT METHODS OF TABULATING CENSUS DATA.

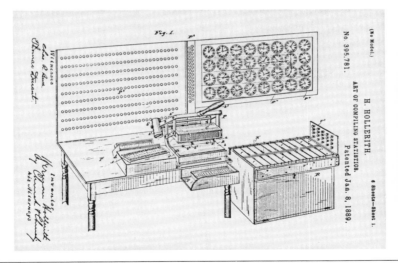

The competition for the best census tabulating system had three entries. William C. Hunt's plan used colored cards. Charles F. Pidgin's system used color-coded paper "chips." Both required hand sorting and counting. The easy winner was Herman Hollerith's electric tabulating machine, which used punch cards.

At nineteen, after graduating from the Columbia University School of Mines, Herman Hollerith began work for the U.S. Census Office. After four years, he left to clerk briefly at the patent office, then went on to work as an engineer. He patented electro-magnetically operated air brakes, and the tabulating machinery which won the census competition. In 1896 he formed the Tabulating Machine Company, one of the organizations that would be brought together to form IBM.

Patent for Hollerith electric tabulating machine. Following a suggestion made by Dr. John Shaw Billings, the medical statistician, Hollerith applied to statistics the punch card information principle of the Jacquard loom.

Hollerith's pantographic punch. When the operator places the pointer in one of the holes on the face of the punch, a corresponding hole is punched in the blank card at the back. Hollerith said his initial inspiration came from watching a railroad conductor use a ticket punch.

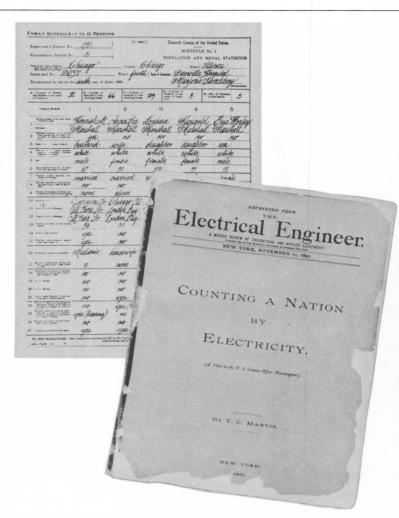

24

For the 1890 population census, a force of
46,804 enumerators completed forms for
each family or household—about
13,000,000 in all. This represented a total
population of 62,979,766.

Hollerith tabulator in use at the
Census Office

"This apparatus works unerringly as the
mills of the gods, but beats them hollow
as to speed."

Electrical Engineer
November 11, 1891

There were occasional mechanical failures, but one of the machine operators candidly recalls: "The trouble was usually that somebody had extracted the mercury from one of the little cups with an eyedropper and squirted it into a spittoon, just to get un-needed rest."

26

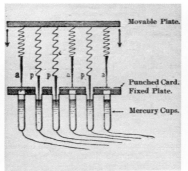

Movable Plate.

Punched Card.
Fixed Plate.

Mercury Cups.

The information collected for each person was punched into cards—each possible answer assigned a specific location on the card.

To read the information on the card into the tabulator, the card was placed in the circuit-closing press. The press had an electrical pin contact for each possible hole location on the card.

Where there was a hole in the card, the pin passed through into a mercury-filled cup below, and completed an electrical circuit.

The system included a semi-automatic sorter wired to the circuit-closing press so that the lid of the appropriate compartment would open in response to specific holes or combinations in the card. After placing the card in the box, the operator closed the lid by hand.

Each completed circuit caused an electro-magnet to advance a counting dial by one number. The tabulator's 40 dials allowed the answers to several questions to be counted simultaneously. At the end of the day, the total on each dial was recorded by hand and the dial set back to zero.

First Russian Census

Following the success of the 1890 U.S. census, Austria and Canada used Hollerith machines for their own censuses. In 1895 Herman Hollerith went to Moscow to sell his tabulating equipment to the Russians, who had never taken a general census.

Hollerith described his selling problem simply: "The principal thing is to get them to have a census."

After months of negotiations, Hollerith had a contract from the Czar, and the Russians had equipment for tabulating the results of their first census in 1897. They did not take another until the first Soviet census of 1926.

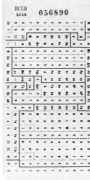

The Church of St. Basil, Moscow

Punch card from the first Russian census

A Russian book promoting the Hollerith system was published at St. Petersburg in 1894, the year before Hollerith's selling trip to Moscow.

Citizens of St. Petersburg, Russia. The census was hotly opposed in many rural areas. Some non-Christians feared it was a Czarist plot to baptize them, while certain Christian sects viewed it as "the net of anti-Christ." Similarly there had been opposition to the first American census of 1790.

For tabulating the Austrian census of 1890, machines based on the Hollerith design were built in Vienna by Otto Schäffler.

Galton: The Measure of Man

In 1891 Sir Francis Galton installed a laboratory in the South Kensington Science Museum in London where visitors could be measured for "height, weight, span, breathing power, quickness of blow, hearing, seeing, colour sense," and other personal attributes.

In developing techniques to analyze and correlate these data, Galton launched the statistical measurement of man.

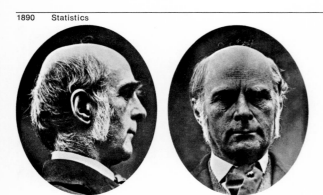

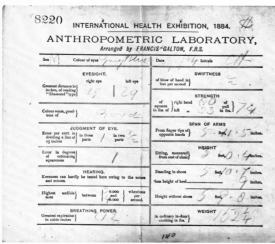

Sir Francis Galton in the full-face and profile views he found most useful for family records and life-history albums

Data card from the Anthropometric Laboratory

Galton, a cousin of Charles Darwin, studied mathematics and medicine but left Kings College, Cambridge without taking a medical degree. Galton was the first to apply statistical methods to biology. In 1869 he published *Hereditary Genius* and went on to develop the science of eugenics. A Fellow of the Royal Society, Galton traveled widely, made innovations in sociology and experimental psychology, and was largely responsible for the introduction of graphical methods into meteorology.

Galton's first Anthropometric Laboratory was part of the 1884 International Health Exhibition in South Kensington Science Museum.

The Anthropometric Laboratory

Over a period of four years, more than nine thousand people paid a fee for the privilege of acquiring and providing Galton with information about their personal capabilities and characterstics. Galton studied the data for ten years. His method of analyzing these data was a basic contribution to statistical science.

To Galton, statistics were "the only tools by which an opening can be cut through the formidable thicket of difficulties that bars the path of those who pursue the Science of Man."

In 1891 the second laboratory was established at the Museum, taking advantage of the regular flow of museum visitors.

Calipers used for anthropometric measurement

Finger Prints

In 1883, French criminologist Alphonse Bertillon had introduced a scientific system of criminal identification based on head and body measurements. It was already widely used by the late eighties, when Sir Francis Galton began collecting fingerprints.

Galton observed that his ideas of correlation would invalidate the Bertillon system. That, plus his proof that no two persons could have identical prints, led him to develop a fingerprint identification system which in time supplanted Bertillon's.

Galton's system was essentially the same method used by the FBI to index fingerprints on punch cards in 1934.

Bertillon as indexed in his own system

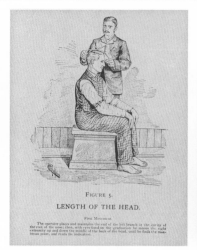

FIGURE 5.
LENGTH OF THE HEAD.

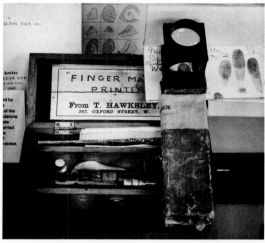

30

FINGER PRINTS

BY
FRANCIS GALTON, F.R.S., ETC.

As a criminologist and an authority on handwriting, Bertillon was a witness for the prosecution at the treason trials of Alfred Dreyfus in 1894 and 1899.

The Bertillon identification system was quickly adopted by police departments in England, Germany, Austria, Russia, Switzerland, and the United States. Here, a New York City police class studies anthropometric identification.

Alphonse Bertillon's *Identification of Criminals,* published in 1889, was the basic text on the subject until the system it taught was superseded by fingerprint identification.

Pocket kit carried by Sir Francis Galton to take on-the-spot fingerprints of his friends, among them Gladstone and Spencer. An inveterate measurer, Galton also carried a device for recording his estimate of the beauty of women he passed in the street.

Fingerprint pattern from Galton's book *Finger Prints*

Although he initially opposed classification by fingerprints, Bertillon in later years incorporated the newer system into his own. Here, he is taking prints in a Paris murder case.

John Gore at Prudential

John Gore was not a professional inventor, but a life insurance actuary, worn out by long hours of calculation. He could see that the Prudential Life Insurance Company, with its extensive statistical departments, desperately needed a mechanical way of handling data.

He drew up rough plans for a perforating machine and a card sorter, and showed them to his brother-in-law, a mechanical engineer. The machines they designed were installed in 1895, and used well into the 1930s.

John K. Gore

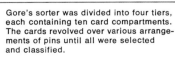

In 1892, at the age of twenty-eight, Gore resigned from a teaching position at the Woodbridge School in New York to join Prudential; by 1912 he had become vice-president of the company.

Gore's multiple-key punch automatically ejected punched cards as new cards were brought into position.

Gore's sorter was divided into four tiers, each containing ten card compartments. The cards revolved over various arrangements of pins until all were selected and classified.

Prudential clerks tabulating insurance premiums with Gore card and sorter. Before inventing his machines, Gore taught clerks to count cards by listening to the sound they made as they were riffled under the thumb.

Marquand's Logic Machine

Princeton professor Allan Marquand had built in the 1880s a machine to solve problems in formal logic. By an arrangement of rods and levers, catgut strings, and spiral springs, the machine displayed all the valid implications of a simple logical proposition.

Marquand later drew up a circuit diagram for operating his machine electrically—the first known design for an electrical logic machine.

Allan Marquand

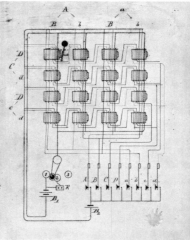

In 1883 Allan Marquand left his position as a tutor of logic at Princeton for a professorship of Art and Archaeology. But he remained interested in logic and devised various machines which could perform logical operations. After abandoning a model based on the logical diagram of John Venn, Marquand constructed a machine of the Jevons type from the wood of a red cedar post "which once formed part of the enclosure of Princeton's oldest homestead."

Marquand's logic machine was an advance over Jevons' logic machine both in ease of operation and the number of terms it could handle.

For example, given the following premises (from Lewis Carroll's *Symbolic Logic*):

*No birds, except ostriches, are 9 feet high;
There are no birds in this aviary that belong to anyone but me;
No ostrich lives on mince pies;
I have no birds less than 9 feet high,*

the machine would give the conclusion— "No bird in this aviary lives on mince pies."

Schematic diagram for an electrical logic machine designed by Marquand, employing single switches to control magnets.

Electrifying Logic

While the Marquand machine worked mechanically and solved only very simple problems in logic, Marquand's correspondence with philosopher Charles S. Peirce produced a deep insight into what machines could be.

In commenting on Marquand's machine, Peirce suggested to Marquand that a system of batteries and switches could be hooked up to solve very difficult problems in formal logic, including most of the theorems of algebra and geometry.

Pastore: Logic on Wheels

The mechanical representation of logic has taken a number of forms. Perhaps the most curious is the contraption invented in 1903 by Annibale Pastore, a professor of philosophy at the University of Genoa, Italy.

Although the machine did not look complicated (or for that matter, logical), its wheels, belts, weights, and differential gears could be hooked up in a bewildering variety of configurations to represent any of 256 syllogistic structures.

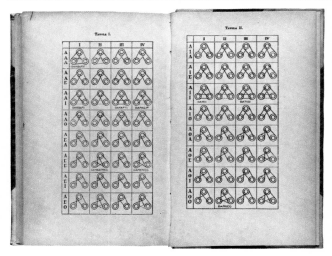

Although Charles Sanders Peirce's father, a well-known mathematician, expected his son to become a scientist, Peirce's career was to span a wide range of disciplines. In 1861 he joined the U.S. Coastal Survey, and he remained thirty years. His duties left him ample time for teaching and research in such diverse fields as logic, physics, criminology, and classical Greek pronunciation. The founder of "pragmatism," he was recognized only after his death as one of the greatest logicians of his time.

A letter from Charles S. Peirce to his former student Allan Marquand contains the first known description of a switching circuit designed to perform logic.

Some of the 256 belt combinations in Pastore's book, *Logica Formale: dedotta della considerazione di modelli meccanici*, which includes the following example of the syllogisms his machine could handle:

Whatever is simple does not dissolve;
The soul does not dissolve;
Therefore, the soul is simple.

After making belt connections corresponding to each premise and the conclusion, the operator would crank wheel A. Since the syllogism is invalid, the wheels would not budge.

Mental Calculators

Under the pressures of rapid business growth some companies began hiring men trained as rapid calculators; and those mental calculators set the standard of performance that the early adding machine makers had to beat.

Once calculating machines were available, it became impractical to use extraordinary mental skills for the rapid solution of arithmetic problems. Ordinary workers, equipped with machines, could do the job.

34

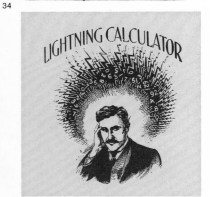

W. K. David, "Practical Rapid Accountant" and author of the *New Practical Lightning Calculator,* promised that "those who diligently examine this book's contents will learn to reckon with phenomenal accuracy and rapidity, and reach a sphere above the ordinary accountant."

Computing was not only tedious and boring—it was unhealthy. A bookkeeper of the period bemoaned his inability to blot out the unending procession of numerical figures that haunted his nights. William Seward Burroughs, inventor of the Burroughs adding machine, was a bank clerk who was forced to change careers because the "monotonous grind of clerical work" had destroyed his health.

Lightning Calculation was a popular form of stage entertainment. Scholars, as well as theater audiences, were intrigued by the phenomenon. The French *Académie des Sciences* appointed a committee to study the mental processes of stage calculators; psychologist Alfred Binet, a member of the committee, wrote a book on the subject.

The rising popularity of calculating aids—and later, calculating machines in the business world—was characteristic of a shift in attitude towards the kind of work people should or could do. The sustained addition of long columns of figures was, Dorr E. Felt noted, "turning men into veritable machines." "Calculating power alone should seem to be the least human of qualities," Oliver Wendell Holmes had written. Harvard President Charles Eliot said, "A man ought not to be employed at a task which a machine can perform."

Léon Bollée

French inventor Léon Bollée built a direct multiplication machine in 1889 when he was eighteen years old. He needed it to help prepare extensive tables of bell dimensions for the family foundry at Le Mans. Bollée invented other calculators and office machines, but in his mature years he was chiefly interested in designing, building, and racing light automobiles. He founded the famous race-track at Le Mans.

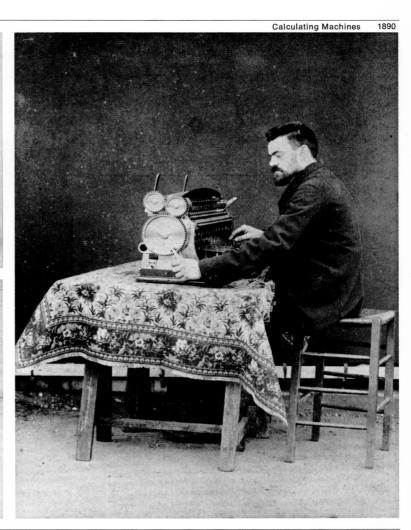

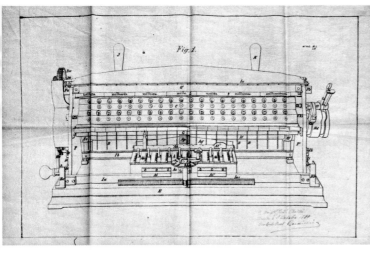

Young Léon Bollée's Multiplier won him a gold medal at the Paris Exposition of 1889.

Patent drawing on linen for Bollée's Multiplier.

Léon Bollée with his Multiplier. Only a few of the massive machines were sold, one of them to the Belgian Ministry of Railroads.

Bollée's long career as an inventor began at the age of thirteen when he patented an unsinkable aquatic bicycle. (An Englishman named Rigby rode it across the English Channel.) He followed this success with others, among them a cash register and a machine which distributed railway tickets. When Wilbur Wright came to France in 1908 to demonstrate his air-plane, Bollée placed his automobile factory at Wright's disposal.

The "Millionaire"

Otto Steiger's "Millionaire" was introduced in 1893. Developed for big business, it was immediately useful in science as well, and was one of the key machines in scientific calculation for about thirty years.

36

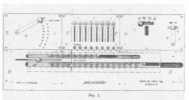

Direct Multipliers

The "Millionaire" worked on the principle of direct multiplication: "one turn of the crank for each figure in the multiplier."

Earlier, Ramon Verea and Léon Bollée had built machines using the same principle, but Steiger's was the first to be commercially successful. Forty-six hundred and fifty-five "Millionaires" were sold between 1894 and 1935, with government agencies the largest customers.

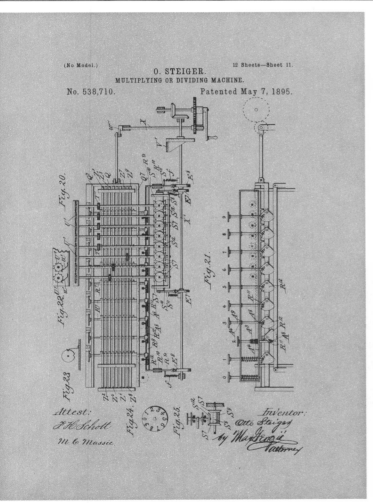

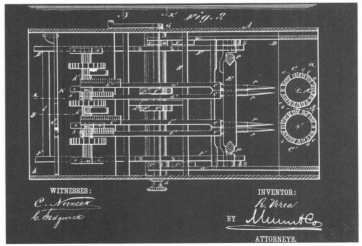

The first direct multiplication machine was built in 1878 by Ramon Verea, a Spaniard living in New York City. Verea told *New York Herald* reporters that he "did not make the machine to either sell its patent or put it into use, but simply to show that it was possible and that a Spaniard can invent as well as an American."

The Comptometer and the Burroughs

The business machine industry really got under way with two American machines marketed in the 1890s: the Comptometer, invented by Dorr E. Felt, and the Burroughs Adding and Listing Machine, invented by William S. Burroughs. They became the most popular accounting machines in the world.

Dorr E. Felt

Never heard of the Comptometer.
TOO BAD.
Having trouble with that trial balance.

Saw our advertisement.
FEELING BETTER.
Has time for other work.

38

Although primarily used for accounting, Comptometers were also valuable in engineering. Around 1895 the U.S. Navy was using them for almost all of the computations required in warship construction.

Felt describes preparation for making his first model: "It was near Thanksgiving Day of 1884, and I decided to use the holiday in construction of the wooden model. I went to the grocer's and selected a box which seemed to me about the right size for the casing.
"It was a macaroni box, so I have always called it the macaroni box model. For keys, I procured some meat skewers from the butcher around the corner and some staples from a hardware store for the key guides and an assortment of elastic bands to be used for springs. When Thanksgiving Day came I got up early and went to work with a few tools, principally a jack knife."

In 1886 Felt persuaded his employer, Robert Tarrant of Chicago, to sponsor his invention; within a year he was a full partner in Tarrant's company.

Popular Accounting Machines

Both Felt and Burroughs believed that accountants and clerks would be best served by machines that could outperform accountants and clerks.

"Now I knew that many account-ants could mentally add four columns of figures at a time," Felt recalled, "so I decided I must beat that in designing my machines."

William S. Burroughs

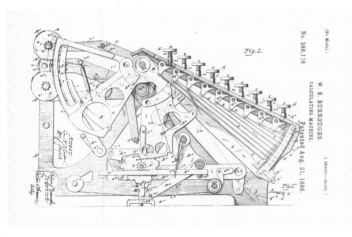

When Burroughs, the son of a mechanic, found it necessary to change occupations at the age of twenty-four, he took up his father's trade. Four years later Burroughs was vice-president of his own company, the American Arithmometer Company.

In 1899 Burroughs manufactured fifty machines, but they proved impossible for almost anyone but Burroughs himself to operate smoothly. (One exception was a field agent who operated his so well that he refused to sell it, preferring to haul it from saloon to saloon in a wheelbarrow, betting drinks on its accuracy.)

The machines were recalled, and Burroughs soon afterward invented a corrective automatic device. One day he went alone to the room where the original fifty were stored, carried them, one by one, to a window, and hurled them to the pavement.

An early Burroughs model. Later ones were encased in glass, and the visibility of the works was used as a selling point.

Accounting Department of J. Lyons & Co., Ltd., London

Calculation by Measurement

A. A. Michelson made crucial measurements of how fast light travels in air, and in water, establishing a quantitative foundation for the wave theory of light.

Michelson needed more precise instruments for calculation and measurement, so he built machines that calculated by measurement—harmonic analyzers.

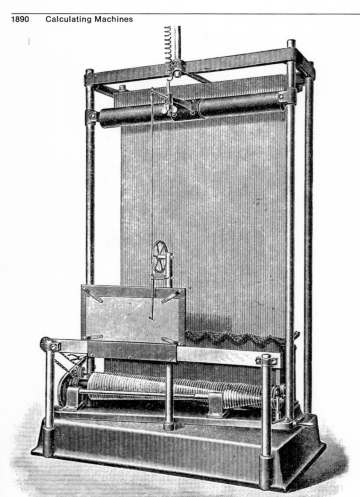

A. A. Michelson

The Michelson-Stratton harmonic analyzer. An earlier, twenty element machine had been built by Michelson based on Lord Kelvin's analyzer of 1876.

Ryerson Physical Laboratories at the University of Chicago. Michelson organized and headed the University's physics department, which became a center for significant experiments on radiation and matter.

Michelson, seated at left, was called by R. A. Millikan "the pioneer in the art of measurement of extraordinarily minute quantities and effects."

When Michelson's application to the U.S. Naval Academy was rejected, he went to President Grant and persuaded him to reverse the decision. After graduating in 1873, he made the first significant improvements on Foucault's method of measuring the velocity of light. As a professor at the Case School in Cleveland, Michelson and his collaborator Edward Morley conducted experiments which are of fundamental importance to the theory of relativity. In 1907 he became the first American scientist to receive the Nobel Prize.

Michelson-Stratton Harmonic Analyzer

Ocean waves, sound waves in a musical chord, the light waves from a star are all complex combinations of simpler waves. Given a complex wave, a harmonic analyzer calculates which simple waves should be combined to produce it.

In 1898 Michelson, working with S. W. Stratton, designed an eighty element harmonic analyzer to study light waves.

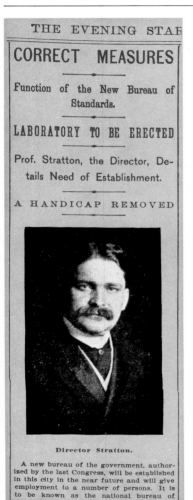

Director Stratton.

A new bureau of the government, authorized by the last Congress, will be established in this city in the near future and will give employment to a number of persons. It is to be known as the national bureau of standards and is to be under the control of

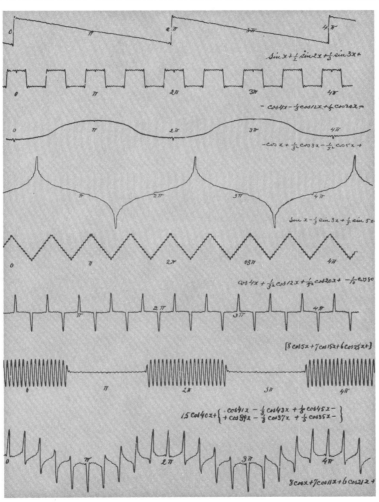

41

Samuel W. Stratton, Michelson's collaborator, prepared a report on the proposed U.S. Bureau of Standards, drafted the bill creating the Bureau, and in 1901 became its first director.

Curves drawn by the eighty element machine. This is an application of Fourier's principle of harmonic analysis. Roughly speaking, he proved that any continuous curve, no matter how irregular, can be approximated by adding together a number of simple regular curves.

The harmonic analyzer being demonstrated by Fred Pearson, Michelson's technician

DAWN OF THE CENTURY

XX CENTURY.

MARCH & TWO STEP

BY E. T. PAULL.

43

The Dynamo and the Virgin

Historian Henry Adams came away from the 1900 Trocadéro Exposition in Paris with a new way of viewing the machine. He concluded that what the Church had been to medieval culture, the dynamo was to ours—an idea he expressed metaphorically by juxtaposing "The Dynamo and the Virgin" in *The Education of Henry Adams.* In it, he wrote, "The new American, like the new European, was the servant of the powerhouse."

Moxon's Master

1900   Logical Automata

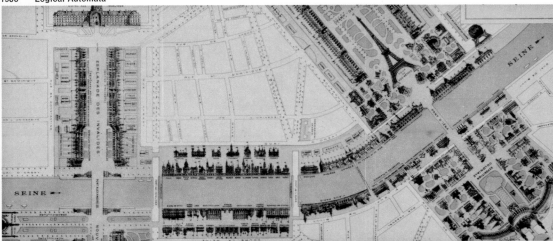

Ambrose Bierce

44

The dynamo and motor exhibit in the Palace of Electricity at the 1900 Trocadéro Exposition

Chartres Cathedral, which was to Adams a symbol of the force of the Virgin: "All the steam in the world could not, like the Virgin, build Chartres."

"If a man during his period of activity is alive, so is a machine when in operation. As an inventor and constructor of machines I know that to be true."

The speaker is Moxon, in *Moxon's Master,* Ambrose Bierce's 1893 story of a thinking machine that turns on its creator. In the story, Moxon constructs a chess-playing automaton. The inventor wins a game, and the enraged machine strangles him.

Alfred Binet: The Scale of Intelligence

French psychologist Alfred Binet insisted, "It would be of little usefulness to study intelligence and character . . . without applying any measure."

In 1908, with Théodore Simon,

Binet published that method of measurement: the Binet-Simon age scale. It introduced the concept of "mental age," and became the basis for standardized psychological intelligence testing.

As such, it underlies the vital role that such testing has come to play in business and industry, in the military, in education, and in medicine.

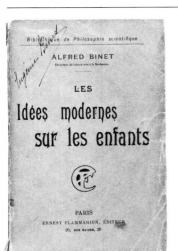

45

As director of the new psychological laboratory at the Sorbonne, Alfred Binet investigated hypnotism, mental fatigue, and "imageless thought." He adapted Galton's statistical approach for a major work, *Etude experimental de l'intelligence,* an investigation of the mental characteristics of his two daughters.

Binet and his collaborators with a time-recording instrument used in tests for mental age

Binet and Simon administered tests to the children of the Rue Grange aux Belles school in Paris.

Théodore Simon recording the results of an anthropometric test

Photograph inscription: *Alfred Binet, sa femme, ses filles et le chien*

Punch Cards for Commerce

Masses of government census data had been quickly and automatically processed by Hollerith machines. Now, large industries began applying the machines to their own data-handling problems. Chief among those industries were railroads, insurance companies, and public utilities.

Herman Hollerith

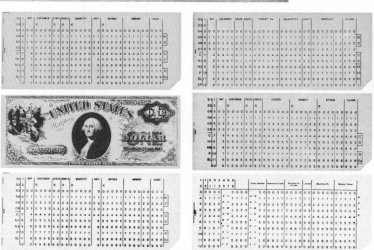

Filing section of the Metropolitan
Life Insurance Company's home office

To accommodate more information for business use, Hollerith increased the size of his punch cards. He chose the size of the dollar bill then in use, and this became a standard.

The amount of information the card could carry was increased in the late 1920s by changing the size and spacing of the holes, without changing the size of the card itself.

As each company extended the use of punch card accounting to new departments, cards were ordered with special layouts and colors.

View taken at the Tabulating Machine Company's factory in Washington, D.C. around the turn of the century

Big Business and Small Facts

Herman Hollerith developed an electric adding mechanism and incorporated it into his tabulating equipment. This made it feasible for railroads to use punch card machines for their waybill statistics— what was shipped, who shipped it, who received it, how much it weighed, the shipping charges, and routes taken.

Insurance companies, with actuarial statistics to correlate and mortality predictions to make, were quick to see the advantages of machine tabulation.

Public utilities had similar problems: keeping track of very large numbers of very small amounts.

Telephone companies had large numbers of transactions to be recorded and billed.

With an electrical integrator, the Hollerith machines could add as well as count.

Railroad freight departments required "a veritable army of clerks."

48

The accounting office at Yale & Towne
Manufacturing Company, March 1908.
Bookkeeping was then primarily an
occupation for men. The English
magazine *Engineering* observed that
manual bookkeeping "would need the
personal attention of someone of
marked ability. But when the data are
punched on cards, the job can be
put in the hands of a girl."

Flexible Plugboard

In 1902 a flexible telephone-type plugboard was installed in the Hollerith tabulator. By rearranging the plugs, the information from any column in the card could be fed into any register. This made it possible to do many applications on one machine without tedious rewiring.

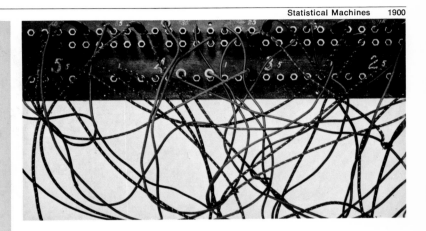

A WONDERFUL MACHINE

KNOWLEDGE OF WHICH WILL HELP YOU PLACE YOUR RATED-UP POLICIES

"THE world do move". The marvels of electricity are everywhere. They help us in the insurance business. The Hollerith Tabulating and Sorting Machines (Figs. 3 and 4) here shown are operated by the electric current. Their results form one of the most interesting and distinct advances in the statistical end of the life insurance business.

Agency Director Langmuir, Los Angeles, writes that he has been greatly aided in placing rated-up policies by the feeling of confidence which the accuracy and reliability of these machines have inspired in him. A word about them will help every man to deliver this business and get it paid for.

There are many different impairments from which men suffer. These impairments affect their rating as life insurance risks in much the same way as nearness to an oil refinery runs up the rating of a fire insurance risk.

Fig. 3.—Machine for Sorting Cards in Groups

The original Hollerith electric tabulating system did not have an adequate method for sorting cards. This became a problem in the 1900 agricultural census, so Hollerith devised an automatic sorter. The first one was a tabletop model with the bins arranged horizontally. Later, when his system was gaining favor commercially, he redesigned the sorter into a sturdier, vertical machine that would not take up too much space in small railroad offices.

After Hollerith incorporated an adding mechanism into his tabulator, the cards were redesigned with columns of numbers—essentially the form used today. This allowed the use of a single 10-key punch for all applications.

Statistical Fallout

At first tabulating machines were used to replace hand methods of accounting. But soon they were applied to tasks that could not have been performed by hand at all—the immediate analysis of costs and sales.

This new machine capability offered marketing advantages in a great variety of businesses—steel mills, department stores, shoe manufacturers, machine shops, packing companies.

1900   Statistical Machines

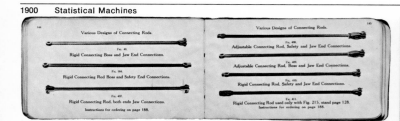

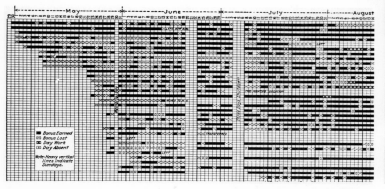

50

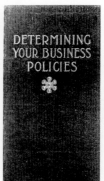

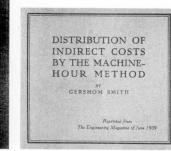

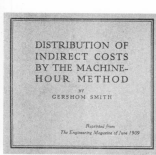

An assortment of connecting rods from the Elliot Frog & Switch Company's catalog indicates subtle product variations that affect cost and pricing.

Photos taken at Pennsylvania Steel's Steelton, Pennsylvania, plant. This was an early electrical cost accounting installation for machine shop work.

Record of work in one room of a worsted mill. Each line represents one worker.

Cost and Sales Analysis

Cost accounting had been largely guesswork. But now, with Hollerith machines, a company could know how much its products actually cost to make, how well they were selling, and where. The machines were used to handle payroll analysis, purchasing records, shipping costs, overhead allocations, inventories.

A sales analysis could be made at the end of each month, each week, each day, and broken down by product line, territory, season, or any other classification management found useful.

Machine shop at Bethlehem Steel Company. The larger the company and the more diversified the product line, the more important accurate analysis became.

Taylorization

Before there could be assembly lines and automation, factory work had to be systematized, just as office work had to be systematized before punch cards and business machines could be used. Systems of work were vigorously introduced to industry by engineer Frederick W. Taylor, apostle of what came to be called "scientific management" or "Taylorization."

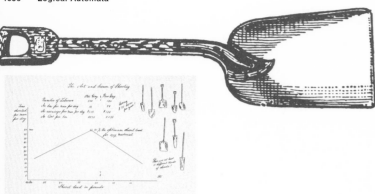

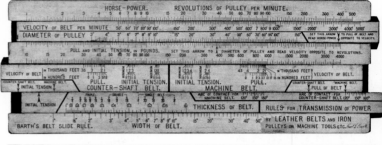

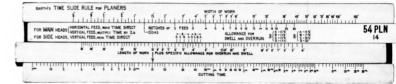

52

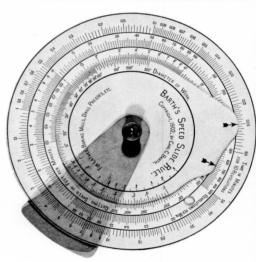

At the Bethlehem Steel Company, Taylor established "the scientific fact that a . . . first-class shoveler will do his largest day's work when he has a shovel load of 21½ pounds." Then, "as a matter of common sense . . . it was necessary to furnish each workman each day with a shovel which would hold just 21½ pounds of the particular material which he was called upon to shovel."

Because of his poor eyesight Frederick Winslow Taylor was forced at the age of eighteen to give up his plans to attend law school. As an engineer for the Midvale Steel Company he conducted experiments intended to maximize the efficiency of metal-cutting. In 1907 he was hired to reorganize Bethlehem Steel Company and improved factory operations considerably. After leaving Bethlehem Steel, Taylor spent his time writing and lecturing on his system of management.

Important instruments in Taylorization were the slide rules developed by Taylor's associate Carl Barth. His first slide rule was only a slight improvement over an earlier one jerry-built by Taylor and others. Barth's refined models gave immediate solutions to "speed and feed" problems of production.

The Copper Man

Author L. Frank Baum shared the turn-of-the-century optimism about machines as a positive force. An admired (but not beloved) character in his famous Oz series was Tik-tok, the clockwork copper man who "was sure to do exactly what he was wound up to do, at all times and in all circumstances."

That a machine "would only do the special kind of thing it had been calculated to do," was regarded as no defect by philosopher Charles S. Peirce, who observed: "We do not want it to do its own business, but ours."

L. Frank Baum, author of the Oz stories and, on the face of it, possible model for mustached Tik-tok, the copper man

Visiting the Pumpkin-Field

a sewing-machine or an automobile. Yet Tik-tok was popular with the people of Oz because he was so trustworthy, reliable and true; he was sure to do exactly what he was wound up to do, at all times and in all circumstances. Perhaps it is better to be a machine that does its duty than a flesh-and-blood person who will not, for a dead truth is better than a live falsehood.

About noon the travelers reached a large field of pumpkins — a vegetable quite appropriate to the yellow country of the Winkies — and some of the pumpkins which grew there were of remarkable size. Just before they entered upon this field they saw three little mounds that looked like graves, with a pretty headstone to each one of them.

"What is this?" asked Dorothy, in wonder.

171

From *The Road to Oz,* by L. Frank Baum, 1909

Astronomical Calculations

The progress of astronomy depends largely upon calculation. Since Napier's invention of logarithms in 1614, astronomers had been relying on logarithmic tables in performing their lengthy computations.

Now, the use of accurate calculating machines began to gain recognition as a superior method.

1900   Calculating Machines

Percival Lowell at center, and observatory staff, 1905

Percival Lowell, brother of poet Amy Lowell, devoted his time to travel, writing, and diplomacy after graduating from Harvard in 1876. Excited by Schiaparelli's discovery of "canals" on Mars, Lowell turned to the study of the planets, founding in 1893 the Lowell Observatory in Flagstaff, Arizona. There he developed his theory about intelligent life on Mars and studied the perturbations of Uranus in the hope of locating an unseen planet beyond Neptune.

Turning the Tables

For calculating in astronomy, machines had proven faster than log tables, but two things prevented their widespread use. They were expensive; and to be used they required another kind of table—the trigonometric functions in their natural, rather than logarithmic form. Such tables were not yet readily available.

Ironically, natural tables were precisely the kind that had been discarded in favor of logarithms.

After three hundred years, their reprinting and refinement made the use of calculating machines in astronomy practical.

Herbert Hall Turner

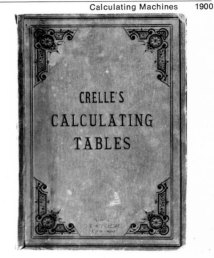

In 1896 W. Jordan reprinted the sine and cosine tables from the three hundred-year-old *Opus Palatinum.* This was, in L. J. Comrie's view, "the first evidence that the introduction of machines had revived the necessity for natural tables." More evidence was to follow.

As chief assistant at the Royal Observatory in Greenwich and later as a professor at Oxford, British astronomer H. H. Turner experimented with photography as an astronomical tool and urged the adoption of daylight saving time. In 1904 he compared the relative speeds of calculating by logarithms and by machine. When both were applied to a common astronomical problem he found that the machine—a Brunsviga calculator— was three times faster.

Crelle's Calculating Tables was the most often used reference for any scientist whose work involved a great many multiplications.

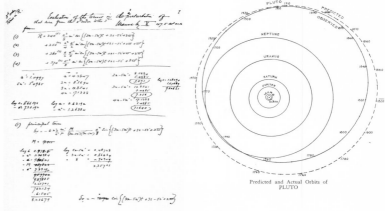

Predicted and Actual Orbits of
PLUTO

56

CALCULATING MACHINE
„THE MILLIONAIRE"

The „**Millionaire**" **with keyboard** is built as 6×6=12 place hand-worked machine and as 8×8=16 and 10×10=20 place machine for motion by hand as well as for motion by electricity.

The keyboard-arrangement makes of the „Millionaire" in cases in which the printing of the items can be dispensed with also a *first class Adding-Machine.* This can chiefly be said of the keyboard-machine with motor-attachment, as this machine can be operated by **the left hand alone.**

Lowell at work

Lowell's calculations for "Planet X"

The calculations for locating Neptune in 1846 had taken Leverrier and Adams five years to do by hand. When Percival Lowell began, in 1905, to search for a "Planet X" beyond Neptune he expected the computations to take more than three years. So he bought a "Millionaire." Even with the machine, the calculations were not completed until 1914.

Lowell died in 1916. After his death, one of his assistants described the planet as "waiting in the sky to be picked out from among the stars." Not until 1930 was it finally sighted. It was named Pluto, with the symbol "℘" after Percival Lowell.

Bjerknes' Weather Mechanics

With the advent of the telegraph came widespread simultaneous weather reports. Forecasters drew weather charts, then made predictions by following the movement of weather conditions across them.

But Norwegian physicist Vilhelm Bjerknes saw that accurate forecasting would depend on mathematical techniques for describing atmospheric behavior. He shifted the focus of weather prediction back to problems of mechanics and physics.

However, the mathematical techniques Bjerknes sought did not appear until Lewis Fry Richardson developed them during World War I. And the technology for implementing them did not appear until the forties.

Vilhelm Bjerknes

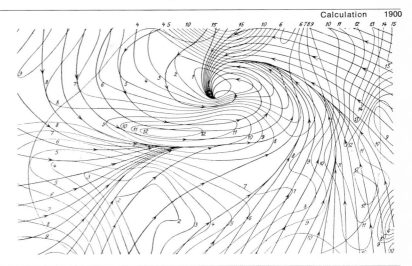

Calculation   1900

A student of Henri Poincaré and Heinrich Hertz, Vilhelm Bjerknes' early training was in the field of hydrodynamics. His work on electric resonance aided in the development of electromagnetic theory of radiation.

Weather Forecasting as a Problem in Mechanics and Physics, in which Bjerknes suggested methods of predicting weather by translating weather data into mathematical terms, was first published in this Austrian meteorological journal of 1904.

During the First World War, Norway was cut off from food imports, as well as weather reports from outside Scandinavia. Convinced that he could improve weather forecasts for Norwegian fishermen and farmers, Bjerknes arranged to increase the number of reporting stations in southern Norway tenfold (from nine to about ninety stations).

Diagram from Bjerknes' *Dynamic Meteorology and Hydrography*, 1910. When the practicality of his work was challenged, Bjerknes replied: "It may require many years to bore a tunnel through a mountain. Many a laborer may not live to see the cut finished. Nevertheless this will not prevent later comers from riding through the tunnel at express-train speed." Bjerknes was still alive when the first numerical forecast based on his principles was made on ENIAC in 1950.

59

Gyroscopic Guidance

The idea of "feedback"—using part of the work done by a machine to control the work it is about to do—had previously been used to regulate the speed of steam engines (with ball governors) or the temperature of furnaces (with thermostats).

But the more complex problem of using feedback to make a vehicle guide itself called for a device that responds to changes of direction. Gyroscopes, which maintain a fixed position in space, proved ideal.

1910   Logical Automata

60

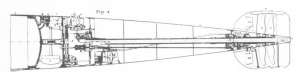

Robert Whitehead

Harper's Weekly of April 29, 1905, reporting on the Whitehead torpedo, said: "It is so controlled by its mechanism that it will carry out its instructions, even varying them according to conditions, almost as though it possessed a brain of its own in its dreaded war-head."

In a steel battleship a magnetic compass could vary as much as 25 degrees with the turning of the turrets. The first practical solution to this guidance problem was the Anschutz Gyroscope patented by Dr. Herman Anschutz-Kaempfe in 1908.

Elmer Sperry and Hannibal Ford in 1911 with a Naval officer on the S.S. *Princess Anne*. They are looking at the first gyroscopic repeater compass.

Resisting Change

Spinning gyroscopes were first mounted in torpedoes. The gyroscope resisted any change in direction by controlling the torpedo's rudder—automatically correcting any deviation from a straight course. The use of this device immediately increased the torpedo's effective range by two and one-half times.

Because the gyroscope's simple linkage to the rudder produced, in effect, overcorrection, the torpedo approached its target in a series of sinuous curves.

By 1910 gyroscopic stabilizing devices had been mounted in ships, and even in an airplane.

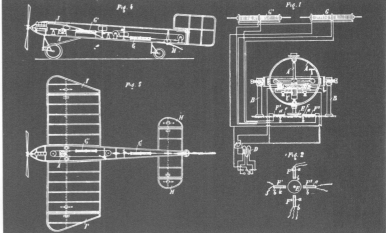

Elmer Ambrose Sperry left Cornell University after one year of casual attendance, and on his twentieth birthday opened a factory for the manufacture of dynamos and arc-lamps of his own design. He first became interested in machinery, he said, during a visit to the Centennial Exhibition in Philadelphia in 1876. Later he turned his attention to coal-mining machinery and diesel engines, and around 1896 began investigating the properties of gyroscopes.

Elmer Sperry first imagined the practical applications of the gyroscope while watching his children play with a toy one. He became a pioneer in the field, developing gyrocompasses, gyrostabilizers, and gyropilots.

Elmer Sperry and Hannibal Ford designed and installed the first gyroscopic stabilizer for an airplane in 1909.

Diagrams from Paul Regnard's 1910 proposal to stabilize an aircraft with a gyroscope

Maintaining an Attitude

When Elmer Sperry installed his gyrostabilizer in an aircraft, he faced a more difficult problem than had torpedo makers in designing gyroscopic guidance. The correction needed to maintain an attitude at one speed might make the air-plane crash at another.

He solved this problem by connecting his device to a wind speed indicator. The gyrostabilizer then automatically modified the force it applied to the ailerons and elevators, depending on the airplane's speed.

62

Elmer Sperry's gyrostabilizer. In 1914, the Aero Club of France announced an international competition, with a prize of 50,000 francs, for a safe airplane.

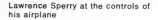

Lawrence Sperry at the controls of his airplane

The Sperry gyrostabilizer installed. The stabilizer, which contained four gyroscopes revolving at 12,000 r.p.m., worked through servomotors to hold the aircraft level.

In 1914, the Aero Club of France announced an international competition, with a prize of 50,000 francs, for a safe plane. Elmer's son, Lawrence Sperry, entered a Curtiss airplane fitted with this gyroscopic stabilizer.

As he flew by the judges' stand, he stood up and waved both arms, while his 170-pound mechanic walked out six feet on the wing. Despite the great imbalance, the airplane stayed absolutely stable. The display was convincing; Sperry won first prize.

Mathematics in Warfare

To illustrate the point that aircraft would fundamentally change the nature of warfare, Frederick William Lanchester made a mathematical analysis of the relation between opposing forces in battle. Lanchester wrote *Aircraft in Warfare* at the beginning of the First World War, but the quantitative consideration of war developed during the Second World War into the science of "Operations Research."

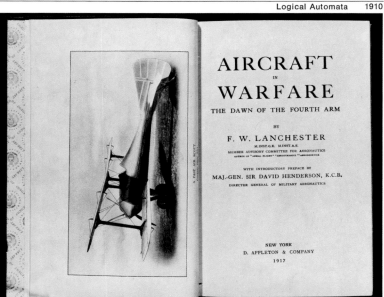

AIRCRAFT
IN
WARFARE

THE DAWN OF THE FOURTH ARM

BY

F. W. LANCHESTER

M.INST.C.E. M.INST.A.E.

MEMBER ADVISORY COMMITTEE FOR AERONAUTICS

AUTHOR OF "AERIAL FLIGHT" "AERODYNAMICS" "AERODONETICS"

WITH INTRODUCTORY PREFACE BY

MAJ.-GEN. SIR DAVID HENDERSON, K.C.B.

DIRECTOR GENERAL OF MILITARY AERONAUTICS

NEW YORK
D. APPLETON & COMPANY
1917

Combined bombsight and electrical mechanical computer. Developed by Hannibal Ford in 1918 for the U.S. Army and Navy, it mechanically determined groundspeed and direction of drift of the airplane with respect to the target. Then it automatically calculated and indicated the correct time to release the bomb.

The famous Norden bombsight, used in World War II, combined the principle of the computing bombsight with that of the airplane gyroscopic stabilizer.

Odhner in Russia

By the time of the Russian Revolution, thirty thousand Odhner calculators had been manufactured. Most of them were sold in Russia.

When the St. Petersburg factory was nationalized, the owners salvaged the documents and blueprints necessary to move the operation to Sweden. There, as Original-Odhner, it has continued to today.

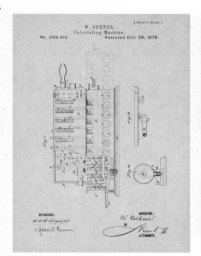

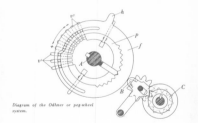

Diagram of the Odhner or peg-wheel system.

The Odhner, invented in 1874 by Swedish engineer W. T. Odhner, is based on the "pinwheel" principle. Since then, many calculating machines, including Brunsviga, Marchant, and Friden, have used the same principle.

The Odhner factory in St. Petersburg

Assembly Lines

In a spectacular 1908 demonstration, three Cadillacs were taken apart, and all the parts mixed up. The cars were then quickly assembled from the mixed-up parts and driven away.

That level of standardization made possible the moving assembly lines installed by Henry Ford some five years later. For assembly lines, the production process was broken down into activities that were repetitive, mechanical, and sequential. This systematization of the factory opened up possibilities of increased mechanization that would, with the development of automatic control devices, lead toward a greater automation of industry.

The magneto assembly line, introduced in 1913 at the Highland Park plant, was Ford's first assembly line.

Overhead conveyors carried wheels, fenders, and running boards to the chassis assembly line.

The "body drop'" end of the assembly line, where the body was lowered onto the chassis

Torres' Theory of Automata

By 1913 mass production was becoming a reality, as manufacturers set up the first moving assembly lines. At the same time, the Spanish inventor Leonardo Torres y Quevedo was demonstrating a theory of automata that looked beyond assembly lines to the industrial use of programmed machines.

To prove that machines could do jobs that seemed to require mental ability, he combined electromechanical calculating techniques with his principles of automata, and showed how a machine could be assembled to perform any desired sequence of arithmetic operations.

1910   Logical Automata

Leonardo Torres y Quevedo

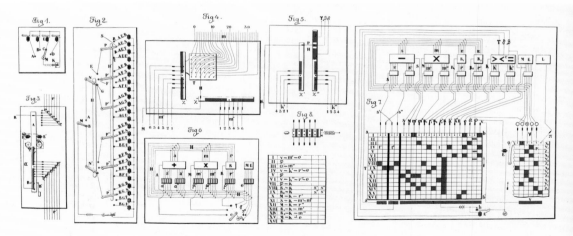

Born in 1852 in Santa Cruz, Spain, Leonardo Torres y Quevedo studied at the Institute of Bilbao and the School of Engineers in Madrid. He developed a remote-control guidance system which successfully steered a boat through Bilbao harbor, dramatizing the extent to which machines could perform tasks formerly reserved for men. He also designed the "Astro-Torres," a dirigible which dispensed with rigid internal structure, and a cable-lift used at Niagara Falls.

Conceptual diagrams for the automatic calculating machine described by Torres to illustrate his thesis about general-purpose automata

Torres' electromechanical arithmometer, exhibited in 1920, was an attempt to realize the theories of his 1913 study of automata. Arithmetic problems were typed in by the operator and the arithmometer caused the typewriter to type out the answers. Torres pointed out that several typewriters could be connected to one arithmometer—a very early suggestion of time-sharing.

Arithmetic, Chess, and Automation

Leonardo Torres built what was probably the first decision-making automaton—a chess-playing machine. Playing an end game with a rook and king against a human opponent's king, the machine would checkmate.

In 1915 Torres, when interviewed by *Scientific American,* said: "The ancient automatons . . . imitate the appearance and movements of living beings, but this has not much practical interest, and what is wanted is a class of apparatus which leaves out the mere visible gestures of man and attempts to accomplish the results which a living person obtains, thus replacing a man by a machine."

Scientific American went on:

"M. Torres claims that . . . at least in theory most or all of the operations of a large establishment could be done by machine, even those which are supposed to need the intervention of a considerable intellectual capacity."

At the Cybernetic Congress in Paris, 1951, Gonzalo Torres y Quevedo matched his father's chess-playing automaton against Norbert Wiener (right).

Torres' Algebraic Machines

Torres built several algebraic equation solvers between 1893 and 1920, each increasingly sophisticated. One of his later designs incorporated a "no-end-axle" machine. A mechanical realization of the additive logarithms of Gauss, it adds numbers while still in their logarithmic form. Four such models were built for Torres at the Laboratoire de Mécanique de la Sorbonne, Paris.

When someone described him as a great mathematician, Torres responded: "Why? Because I invented an algebraic machine? No, that is not so; the machine knows much more mathematics than I."

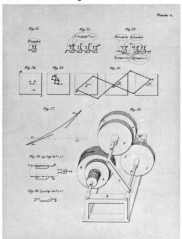

68

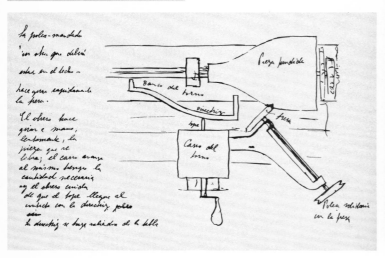

Diagram of the internal operation of the "no-end-axle" machine. It was used to find the roots of polynomials.

Sketch made by Torres in 1902, showing how any desired mathematical function could be represented mechanically

Torres' 1893 algebraic equation solver. It employed disks, graduated in a logarithmic scale, interlinked with gears. With the machine, one could find roots of polynomials up to the ninth degree.

Torres built a simpler and more refined model in 1905.

A Great Brass Brain

In 1914 *Scientific American* hailed the arrival of "a great brass brain," a tide predictor designed for the U.S. Coast and Geodetic Survey by E. G. Fischer and R. A. Harris.

The new machine, fifteen years in the making, was a vast improvement over its predecessor; it computed on the basis of thirty-seven tidal components, displaying the results directly on dials.

Tides are now predicted by a large electronic computer; this task takes only four hours a year.

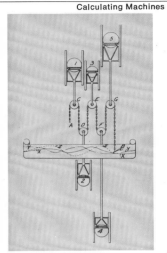

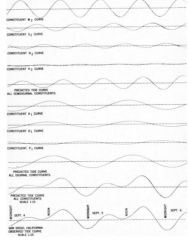

E. G. Fischer, codesigner of the great brass brain, checking its visual display dials

How the curves are summed up mechanically

Simple waves are added together to make a tide prediction curve. The prediction curve, at bottom, is such a sum with the actual tidal curve below it.

During the First World War, ships used tidal data to maneuver into shallow water —eluding the German U-boats. In response, the Germans quickly built this tide predictor. It was completed in 1916, and installed in the Imperial Observatory.

Pearson's Battle For Biometrics

"The battle has lasted for nearly twenty years," Karl Pearson could report in 1920, "but there are many signs now that the old hostility is over and the new methods are being everywhere accepted."

The "new methods" were those of Pearson's Biometric School, which took statistical techniques originally devised for biology and applied them to such other fields as medicine, anthropology, psychology, and astronomy.

1910   Calculating and Statistical Machines

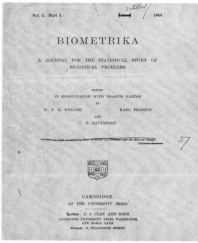

Karl Pearson in his laboratory, 1910.
His Brunsviga calculator is on the desk.

As a student at Kings College, Cambridge, and later at Heidelberg and Berlin, Karl Pearson showed an interest in mathematics, political science, metaphysics, and medieval languages. In 1882 he began to practice law and to write on a wide variety of subjects, but within two years he had been persuaded to accept a professorship in mathematics at University College London. There he studied the applications of statistical methods to biological systems, particularly the Dar-winian theory, of which he was an enthusiastic supporter. Also interested in eugenics, he wrote an extensive biography of his friend Sir Francis Galton.

Corrected proof of the first issue of *Biometrika,* a journal of applied statistics founded by Pearson and his associates, and the manuscript of W. P. Elderton's *Tables for Testing the Goodness of Fit of Theory to Observation,* first published in *Biometrika* in 1902.

The Science of Statistics

Pearson was a major force in making a science of statistics. He established the first "Goodness of Fit" tables—a method of testing the validity of a sample by showing whether the results "can be reasonably supposed to have arisen from random sampling."

An aggressive popularizer of statistical science, Pearson introduced his techniques into a variety of disciplines. His Biometric Laboratory at University College London, was largely responsible for making statistics a branch of applied mathematics.

DEPARTMENT OF APPLIED STATISTICS
(COMPUTING SECTION)
UNIVERSITY OF LONDON, UNIVERSITY COLLEGE

TRACTS FOR COMPUTERS

EDITED BY KARL PEARSON, F.R.S.

No. I

Tables of the Digamma and Trigamma Functions

BY ELEANOR PAIRMAN, M.A.

CAMBRIDGE UNIVERSITY PRESS
C. F. CLAY, MANAGER
LONDON: FETTER LANE, E.C. 4
ALSO
H. K. LEWIS & CO., LTD., 136, Gower Street, London, W.C. 1
WILLIAM WESLEY & SON, 28, Essex Street, London, W.C. 2
Chicago: University of Chicago Press
Bombay, Calcutta, Madras: Macmillan & Co., Limited
Toronto: J. M. Dent & Sons, Limited
Tokyo: The Maruzen-Kabushiki-Kaisha
1919

During the First World War, Pearson's Biometric Laboratory computed shipping statistics, tables of stresses for airplane propellers, and trajectories for antiaircraft guns. Methods and processes had to be invented as need arose. After the war, Pearson published *Tracts for Computers* (by "computers" Pearson meant people who perform computations), explaining that the computers on his staff had "been struck by the absence of any simple textbook for the use of computers."

At the hour when the Armistice was signed in 1918, Pearson was teaching a wounded New Zealander, L. J. Comrie, to use a Brunsviga calculator. Comrie later became the foremost propagandist for the use of business machines in scientific calculation.

Manuscript of the introduction to Pearson's 1914 *Tables for Statisticians and Biometricians*, an indispensable handbook for the statistician.

Powers Printing Tabulator

The Census Bureau, which had been renting tabulating machines from Herman Hollerith since 1890, had decided in 1905 to build its own. It hired James Powers to help with the development.

At the Bureau, Powers invented an electric punching machine, but in 1911 he resigned to manufacture his own machines. His most successful invention was a tabulator that automatically printed its results.

1910   Statistical Machines

72

Russian-born James Powers had completed his technical schooling by the time he came to America at the age of eighteen. Before beginning work on tabulating machines he had patented several photographic inventions, a bread box, and a cutting toothpick.

The Bureau of the Census won a grand prize at the 1915 Pan-Pacific Exposition for an exhibit of methods and machines used in handling statistics.

Powers printing tabulator in use at the New York Life Insurance Company. These frames are from a movie made at the company's home office in 1915.

Powers card-sorting machine which sensed cards mechanically, rather than electrically as in the Hollerith machines.

Some of the three hundred punching machines Powers built for the 1910 Census

Powers electric card-punching machine.
With this machine, the information was
keyed in, then punched all at once
into the card.

Facts and Government

To mobilize for World War I, the Wilson administration was given extraordinary powers. Public agencies were set up to control transportation and communication systems and to regulate production and distribution from factories, mines, and farms.

To operate effectively, the agencies needed rapid access to lots of information —and large numbers of tabulating machines were installed.

1910   Statistical Machines

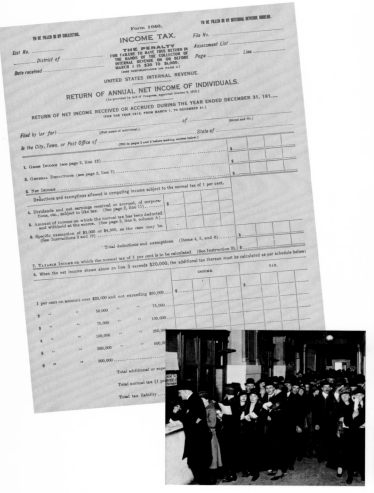

Americans lining up in 1913 to pay their first income tax

Scores of subordinate departments, boards, and committees gathered information, tabulated and sorted it by machine, and fed it to the War Industries Board.

More Government, More Facts

Most powerful of wartime control agencies was the War Industries Board, headed by noted financier Bernard M. Baruch. It was charged with regulating the production of virtually all American industries and controlling the distribution of their goods.

Bernard Baruch (seated at right) and the War Industries Board. The Board worked with cooperating committees from various industries. There were committees on cars, baby buggies, candy, elevators, and foundry supplies. There was the Ice Committee and the Biscuits and Crackers Committee, the Pocket Knife Committee, and the Phonograph Committee. There were committees on seeds and envelopes and harness leather, and an all-important committee on wheels.

Un-Uniform Soldiers

One of the problems of wartime mobilization is putting draftees into jobs (and uniforms) that fit.

The U.S. Army in World War I conducted the first large-scale application of psychological testing to the problem of efficiently using human resources.

The process supplied masses of data, organized by punch card sorter, which were invaluable not only for the immediate problem but for later analysis.

1910   Statistical Machines

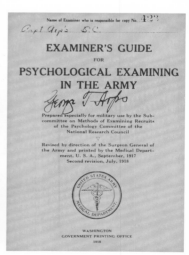

Of the Army's first company of enlisted psychologists, these men received white headbands, the highest rating.

Pictorial competition test being administered to a recruit

Edward L. Thorndike, a pioneer in applying quantification to the psychology of learning, headed the Army's wartime Statistical Unit. Such projects served, in effect, as laboratories for experienced psychologists, and as a training ground for young psychologists.

Psychological Testing

The Army's *Alpha* and *Beta* tests were designed to discover special skills and leadership capabilities. They were also designed to identify recruits likely to be useless, or even dangerous, in battle. *Alpha* was given to literates, *Beta* to illiterates. The results, coded on cards, were used to fill such specialized personnel needs as 105 scene painters for camouflage work, or 600 chauffeurs who spoke French.

Measuring Men

At the end of the war, the Medical Department of the Army measured 100,000 men to secure data for the making of new uniforms. The assembled information provided the first reliable picture of the size and weight of American men. Men from North Dakota had the largest chests, those from Alaska were the heaviest, and, true to stereotype, Texans were the tallest.

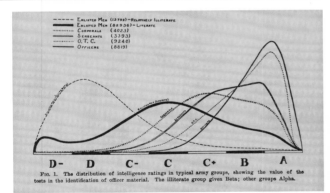

MEASUREMENTS—ALL METRIC

0. Weight	145.0
1. Height, standing (stature)	160.1
2. Span (maximum, between finger and tips of outstretched arms)	161.0
3. Height, sitting	86.0
4. Height of sternal notch	132.0
5. Height of pubis	76.8
6. Transverse diam. of shoulders at level of head of humeri	42.7
7. Transverse diam. of hips, level of crests of ilia	29.8
8. Transverse diam. of chest at level of nipples, arms elevated and flexed	28.8
9. Ant.-post. diam. chest; level of nipples	21.0
10. Second dorsal vertebra to styloid process of ulna (elbow bent, horizontal)	68.0
11. Circumference of neck, level of laryngeal prominence, perpendicular to axis of neck	38.0
12. Circumference of chest, level of nipples	88.7
13. Circumference of waist, level of umbilicus	83.0
14. Circumference of thigh (maximum)	55.2
15. Circumference of leg, just above patella	37.0
16. Circumference of knee, level of patella	35.0
17. Circumference of leg, just below level of tuberosity of tibia	31.9
18. Circumference of calf (maximum)	34.7
19. Inside length of leg, from gluteal fold to tip of internal malleolus of tibia	64.0
20. Size of shoe worn if fitted since July 15, 1919, under par. 14, S. R. 2s	8-C
Number of measurer	6
21. Height of knee	43.4
22. Length of forearm	23.0

DOTTED LINES INDICATE DIAMETERS
FULL LINES INDICATE CIRCUMFERENCES
NOS. 2, 3, AND 10 NOT SHOWN ON FIGURE

Legend:
- ENLISTED MEN (1379²)—RELATIVELY ILLITERATE
- ENLISTED MEN (82936)—LITERATE
- CORPORALS (4023)
- SERGEANTS (3393)
- O.T.C. (9240)
- OFFICERS (8819)

D- D C- C C+ B A

Fig. 1. The distribution of intelligence ratings in typical army groups, showing the value of the tests in the identification of officer material. The illiterate group given Beta; other groups Alpha.

The Committee on Developing Methods of Psychological Examination, 1917, headed by Major Robert M. Yerkes

Psychologist H. H. Goddard, recalling the Committee's work, wrote: "We worked out material for the Test of Common Sense—including such problems as whether cows have horns because we need horn in the art of making combs, because they add beauty and dignity to the appearance of the cow, for protection, or because our grandfathers used them for powder horns. Also a series of disarranged sentences such as 'Hell to with Bill Kaiser.' "

Despite the information collected at the end of World War I, the sizing of uniforms in the next war was not totally successful. The Quartermaster Corps in World War II ignored the "normal curve" of distribution and provided the fliers at Wright Patterson Field with coveralls in ten sizes —10 percent of each size.

Aberdeen

The introduction of new kinds of artillery and ammunition in World War I demanded new and more accurate ballistics tables.

At the Army Proving Ground, at Aberdeen, Maryland, a group of university mathematicians was hastily assembled to apply scientific techniques to the preparation of precise gunnery tables.

1910   Calculating Machines

78

Aberdeen Proving Ground, 1917

Antiaircraft gun being tested at Aberdeen. Norbert Wiener recalled that "the old-fashioned methods of computing range tables had . . . broken down completely in the new and very exacting field of anti-aircraft fire."

Young mathematicians at Aberdeen; Norbert Wiener is at right.

The typescript page is from Wiener's *Ex-Prodigy*, in which he observes that "for many years after the . . . war, the overwhelming majority of significant American mathematicians was to be found among those who had gone through the discipline of the Proving Ground." Wiener also remembers: "When we were not working on the noisy hand-computing machines which we knew as 'crashers' we were playing bridge together . . . using the same computing machines to record our scores."

Improving the Proving Ground

Early in the war the Germans had test fired a large naval gun. In the thinner air at high altitudes, the shell went twice as far as they had expected.

The U.S. Army, faced with the same problems, brought in astronomer F. R. Moulton, who developed a theory, and Princeton mathematician Oswald Veblen, who paralleled the theory with new wind tunnel and proving ground experiments. Their work dramatically improved the precision of gunnery tables.

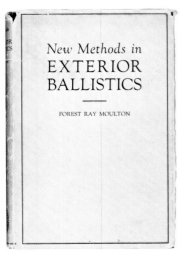

F. R. Moulton, Chief of Ordnance in World War I

In 1969, *Ordnance* magazine, tracing the background of the electronic calculator ENIAC, wrote: "The equations of motion by F. R. Moulton, giving a numerical solution to projectile trajectories, uprooted the forest of empiricism and planted in its place the seed of massive calculations."

Major Oswald Veblen was responsible for preparing tables of gun and ammunition types, map data, powder charge temperature, and such meteorological information as air temperature and density and wind velocity.

High-speed photographs of trajectories taken at C. Cranz's Berlin laboratory before the First World War.

Working with Cranz gave Colonel H. H. Zornig the idea of setting up the Ballistic Research Laboratories at Aberdeen in 1938.

A Weather Forecast-Factory

Nearly forty years before electronic computers, Lewis Fry Richardson imagined a "forecast-factory" in which thousands of mathematicians raced the weather around the globe.

From this factory he developed the basis of a model for numerical weather prediction.
The model was fundamentally the same as that used today by computers.

Lewis Fry Richardson and his wife, Dorothy

"After so much hard reasoning, may one play with a fantasy? Imagine a large hall like a theatre. . . . The walls of this chamber are painted to form a map of the globe. The ceiling represents the north polar regions, England is in the gallery, the tropics in the upper circle, Australia on the dress circle and the antarctic in the pit. A myriad of computers are at work upon the weather of the part of the map where each sits, but each computer attends only to one equation or part of an equation . . . Numerous little "night signs" display the instantaneous values so that neighboring computers can read them. Each number is thus displayed in three adjacent zones so as to maintain communication to the North and South on the map. From the floor of the pit a tall pillar rises to half the height of the hall. It carries a large pulpit on its top. In this sits the man in charge of the whole theatre . . . One of his duties is to maintain a uniform speed of progress in all parts of the globe. In this respect he is like the conductor of an orchestra in which the instruments are slide-rules and calculating machines. But instead of waving a baton he turns a beam of rosy light upon those who are behindhand."

A VISIT TO THE MOONBEAM LAMP Co

The above photo is of the lamp testing (mechanical tests) dept. The M.L.Co. believes in combining pleasure with duty. The vacuum in their lamps is so strong that should one burst the explosion is most violent. The batsman wears mail armour. One officer ⊕ is employed to remove any which fall.

Drawing tungsten wire. The wire itself is too fine to be seen.

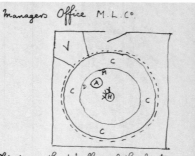

Managers Office M.L.Co.

This is on the top floor of the factory. The telephones, and the periscopes which look into the various floors of the factory are attached to the central pillar H. The chair is marked A. The desk CCCC is ring-shaped.

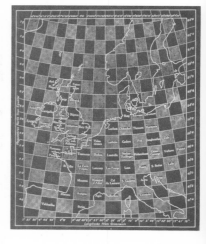

Richardson, the youngest son of an English tanner, became interested in weather while superintendent of Eskdalemuir Observatory. Although he had no formal mathematical training, he was able to teach himself enough to solve his research problems. A man of deep conviction, he worked with the Friends' Ambulance Unit during World War I. In 1929 he began an intensive study of the causes of war, turning down a longed-for professorship in order to continue his researches.

Richardson's drawings of the "Moonbeam Lamp Company." (One of his first jobs had been with the Sunbeam Lamp Company, Gateshead, England.)

The layout of the manager's office resembles the hall that Richardson later imagined for his forecast-factory.

In *Weather Prediction by Numerical Process,* Richardson pictured 64,000 mathematicians working in a great hall to forecast the weather. Outside were playing fields, houses, mountains and lakes, for Richardson believed that "those who compute the weather should breathe of it freely."

An arrangement of meteorological stations designed by Richardson "to fit with the chief mechanical properties of the atmosphere"

Richardson's Dream

Richardson rejected the possibility of ever predicting weather accurately on the basis of previous weather conditions. Instead, he turned to a numerical model in which equations simulated the physical system.

"Perhaps some day in the dim future," he wrote, "it will be possible to advance the computations faster than the weather advances and at a cost less than the saving to mankind due to the information gained. But that is a dream."

In his book *The Mathematical Psychology of War,* Richardson tried to apply mathematical concepts and techniques to the examination of human conflict.

Lewis Fry Richardson: with his sister; as a Cambridge student; at age 23 with his wife; in 1936, at age 55.

A garden party held in Richardson's honor when he became Director of Eskdalemuir Observatory, 1913

Richardson was a Quaker and conscientious objector. His wife recalled, "There came a time of heartbreak when those most interested in his 'upper air' researches proved to be the 'poison gas' experts. Lewis stopped his meteorological researches, destroying such as had not been published. What this cost him none will ever know!"

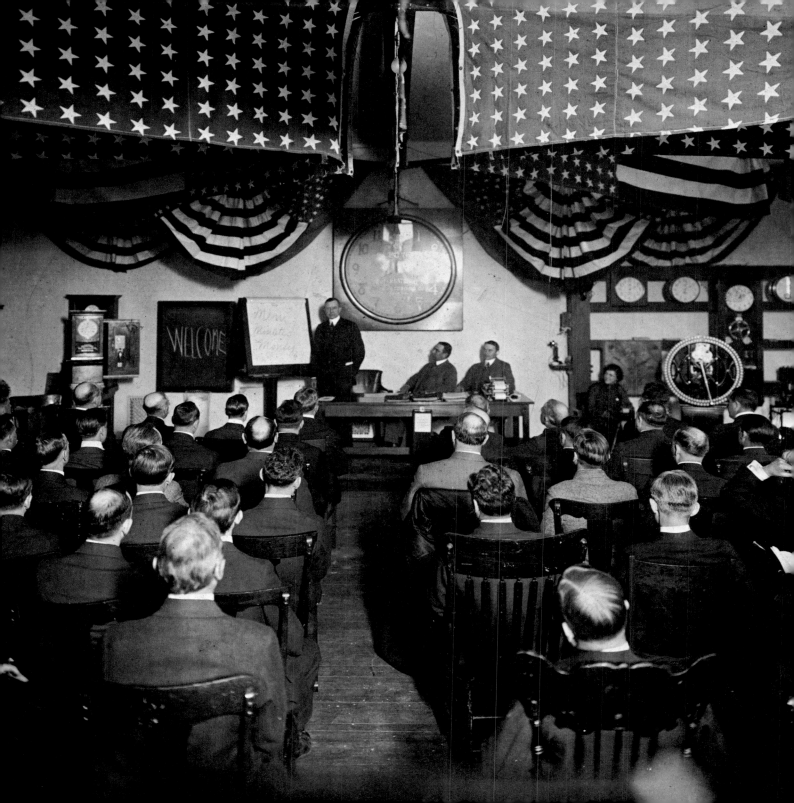

1920

83

Thomas J. Watson Sr. giving his "Men-Minutes-Money" speech at a sales convention in Endicott, New York

Bush's Profile Tracer

"Way back about 1912 I built a machine which was intended to simplify surveying practice. One could run it over a tract of land and it would automatically draw the profile . . . That machine contained an integrator, and it also contained a servo-mechanism. Quite a few years later both of these elements became useful and the first rather crude differential analyzer was built."

Vannevar Bush

Vannevar Bush with his profile tracer

In 1912 Vannevar Bush was awarded a patent for his profile tracer. The machine's movement was transmitted by gears to a disk that in turn controlled the marker and record sheet. The marker drew a vertical profile showing the elevations of all points along the line traveled.

The Product Integraph

Vannevar Bush spent several years in the mid-twenties, developing the "product integraph." It could semiautomatically solve in hours problems in advanced electrical theory that would otherwise have required months of hand computation.

In describing his machine, Bush said, "The product integraph might be called an adding-machine carried to an extreme in its design. Where workers in the business world are ordinarily satisfied with addition, subtraction, multiplication and division of numbers, the engineer deals with curves and graphs which represent for him the past, present and future of the things in which he deals."

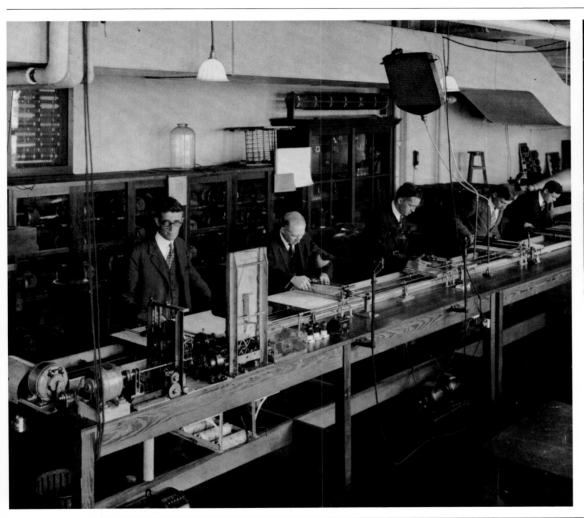

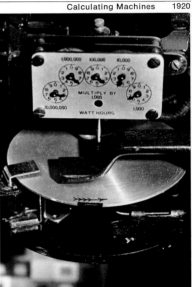

Vannevar Bush with F. G. Kear, H. L. Hazen, H. R. Stewart, and F. D. Gage using the product integraph. The equations to be integrated were plotted by hand on sheets of paper. These were then slowly passed under pointers, and operators stationed along the machine kept the pointers on the curves.

The foundation of the integraph was a watt-hour meter, similar to those installed in homes.

As the pointers of the integraph moved up and down, they controlled the power flowing through the meter. The meter directed a motor to drive a pencil across a slowly moving paper. The curve that was traced expressed the result sought.

L. J. Comrie and Scientific Calculation

British astronomer L. J. Comrie felt that "the time is ripe for the undermining of the old wooden structure of Logarithms, and the rebuilding of our methods in the ferro-concrete of machines."

Comrie was the first to apply Hollerith machines to the physical sciences, computing the predicted motion of the moon from 1935 to the year 2000.

He was the most influential champion of the calculating machine's use in science.

1920   Calculating Machines

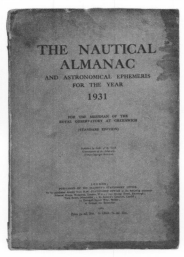

86

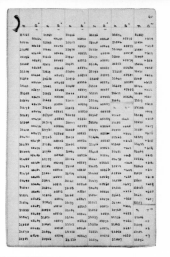

THE HOLLERITH AND POWERS
TABULATING MACHINES

By L. J. COMRIE, M.A., Ph.D.

Being a Lecture delivered under the auspices of
the Office Machinery Users' Association
at the London School of Economics
on 1929 November 20

Reprinted from the
Office Machinery Users' Association Transactions
1929-30

INVERSE INTERPOLATION
AND
SCIENTIFIC APPLICATIONS OF
THE NATIONAL ACCOUNTING
MACHINE

BY
L. J. COMRIE

REPRINTED FROM THE SUPPLEMENT TO THE JOURNAL
OF THE ROYAL STATISTICAL SOCIETY,
Vol. III, No. 2, 1936.

(PRINTED FOR PRIVATE CIRCULATION)
1936.

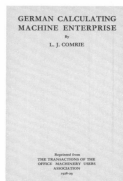

GERMAN CALCULATING
MACHINE ENTERPRISE
By
L. J. COMRIE

Reprinted from
THE TRANSACTIONS OF THE
OFFICE MACHINERY USERS
ASSOCIATION
1928-29

Computing by
Calculating Machines

By Dr. L. J. COMRIE

Leslie John Comrie had an M.A. in chemistry when, after serving in World War I, he decided to become an astronomer. He left the Nautical Almanac Office in 1936 to found The Scientific Computing Service and to publish improved versions of mathematical tables.

As Deputy Superintendent of the British Nautical Almanac office, Comrie proved the effectiveness of mechanical methods of computation.

Comrie used a commercial bookkeeping machine as a difference engine to produce this table of Bessel functions.

Comrie encouraged the development of computational techniques particularly suited to commercial accounting machines. He felt, "An intelligent person can get more out of a machine than a person who simply uses it to do what could be done with a pencil and paper."

At a scientific gathering on Lake Balaton, Hungary, in 1930, Comrie (center) meets with German astronomer K. H. W. Kruse and noted mathematical table maker J. Peters.

Comrie's introduction of machine calculation was first reflected in *The Nautical Almanac* for 1931.

Motions of the Moon

Comrie took data from E. W. Brown's *Tables of the Moon* and punched them onto half a million cards. Something like a hundred million figures were added over seven months.
Later, Comrie wrote, "I showed this to Brown in the summer of 1928; he had done a great deal of this synthesis himself by hand, and I shall ever remember his ecstasies of rapture as he saw his figures being added at the rate of 20 or 30 a second."

"Modern Babbage Machines"

At the end of the 1920s, the National-Ellis 3000 and the Burroughs Class 16 (both capable of accumulating numbers in several registers) came on the market. Comrie conceived a way of using them as difference engines for checking and producing nautical tables.
He hailed these new machines as "Modern Babbage Machines"— "a machine which does what the Babbage difference engine was *intended* to do."

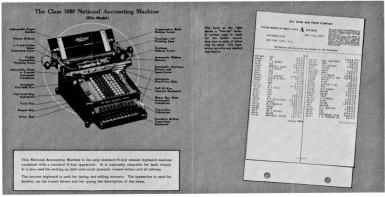

A. J. Thompson's integrating and differencing machine. Thompson connected four Triumphator business machines so that the number on the register of any machine is transferred to the setting levers of the machine immediately below.

Once commercial machines could be used as difference engines, this type of lash-up was no longer necessary.

Salesman's cross section of a Burroughs Class 16 calculator

Corn and Correlation

Henry A. Wallace was concerned with influences on corn yield—rainfall and temperature and soil.

At a time when "corn shows" were giving prizes for the appearance of corn, Wallace, sixteen years old, test-planted five acres and found no correlation between beauty and high yield. And yield was what counted. "What's looks to a hog?" he asked, writing in *Wallaces' Farmer* in 1907. To determine mathematical correlations takes calculation. Wallace studied mathematical statistics, and devised some shortcuts in the use of machines. In 1924 he taught those methods to a group of professors at Iowa State University, which was to become a center for biometric research. By 1927 the University had set up the "Mathematics Statistical Service," the first such installation in America.

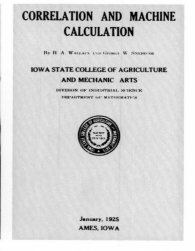

88

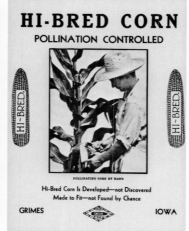

Henry Wallace began, in 1913, to develop methods and shortcuts in computing correlations with calculating machines. For his 1924 Iowa State lectures, he even hauled punch card machines back and forth from an insurance company in Des Moines.

Henry Wallace in 1913 showing his grandfather the results of his first experiments in hybrid corn—a cross between Boone County White and Early Wisconsin. Wallace's interest in genetics began early. He remembered: "George Washington Carver often took me with him on botany expeditions, and it was he who first introduced me to the mysteries of plant fertilization."

In 1924 Wallace developed a funny looking corn named Copper Cross, and won a gold medal at the Iowa Corn Yield Test—the first time a hybrid had beaten standard open pollinated strains. To market his product Wallace founded the Hi-Bred Corn Company, whose slogan was "Developed not Discovered, Made to Fit—not Found by Chance." By 1942 the Corn Belt had been revolutionized; 98 percent of all corn planted in Iowa was hybrid.

In 1925 Wallace collaborated with mathematician George W. Snedecor to publish *Correlation and Machine Calculation.*

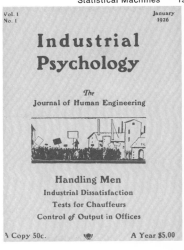

Vol. I
No. 1

January
1926

Industrial Psychology

The
Journal of Human Engineering

Handling Men

Industrial Dissatisfaction

Tests for Chauffeurs

Control *of* Output in Offices

A Copy 50c.   A Year $5.00

The editorship of *Wallaces' Farmer* was a stepping-stone into politics for three generations of Wallaces. Henry A. Wallace was editor from 1924 until 1933, when he became Secretary of Agriculture. He became Vice President under Roosevelt in 1940, and in 1948 campaigned on the Progressive ticket for the Presidency.

Clark Hull reported that in statistics, ''time-consuming drudgery, as well as the arithmetical errors, lie in the lengthy preliminary manipulation of the data.'' To solve this problem he developed his ''automatic correlation calculating machine'' which was run by paper tape.

Stuart Dodd, a research fellow in psychology at Princeton, built a working model of a machine to calculate correlations automatically, and demonstrated it at the Franklin Institute in 1926.

Based on the results of the early model, he built three ''Dodd Correlators.'' He sold one each to Harvard, the University of California, and the University of Chicago.

Interest in using correlations to increase the accuracy of predictions spread to fields other than agriculture, such as industrial psychology, and vocational guidance.

Thomas J. Watson Sr. and the Business of Machines

As president of IBM, Thomas J. Watson Sr. was the first to see an enormous potential in punch card machines. He instilled his vision in his salesmen and engineers, and communicated it to a rapidly increasing number of users in business, government, science, and education.

His success along those lines made him a significant personal force in the invention and popularization of information-handling machines.

1920   Statistical Machines

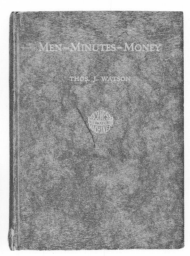

C-T-R president Thomas J. Watson (right) with chairman of the board George Fairchild

Thomas J. Watson Sr. was born in Campbell, New York, in 1874. At eighteen he began work as a grocery bookkeeper, then sold sewing machines and musical instruments. He joined the National Cash Register Company, and within nine years advanced to general sales manager. Three months short of his fortieth birthday, Watson left the company to become head of the Computing-Tabulating-Recording Company—the company that was to become IBM.

"There is no saturation point in education," Watson believed, and he started salesmen's classes in 1916. Next, he added customer engineer's schools, and in 1920 invited all his executives to school at Shawnee-on-Delaware.

One customer, General Electric, knowing that the sales classes provided a sound understanding of the machines, enrolled five of their administrators in 1934. Soon other customers became interested in this

training, and in 1936 the Customer Executive School was officially launched. In a short, intensive course, customers learned to utilize their IBM equipment more effectively. By 1970 some 200,000 had done so.

This collection of speeches takes its title from one of Watson's favorites. His oratory was peppered with morale-boosting slogans to give his salesmen confidence in themselves and their products.

Men, Minutes, Money

In 1911 industrialist Charles Flint combined Herman Hollerith's Tabulating Machine Company with three other small companies to form the Computing-Tabulating-Recording Company. Three years later he chose Thomas J. Watson to head it. In ten years, C-T-R's business had tripled, and Watson optimistically reorganized the company as the International Business Machines Corporation.

Watson had a great faith in the future of his business. When, in 1929, he read a report that only 2 percent of the accounting in the United States was being done by machinery, Watson said: "Think of that! I haven't been able to get that statement out of my mind since I read it. Two per cent! Think of the field we have to work upon."

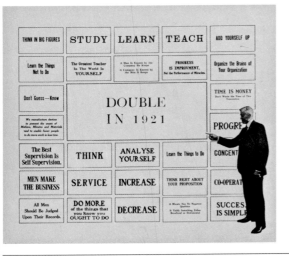

The staff of the Tabulating Machine Company on the roof of their offices at 50 Broad Street, New York City

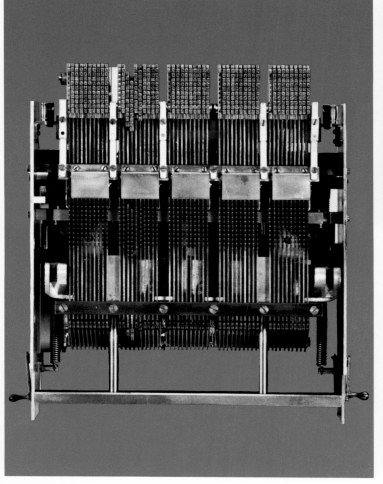

92

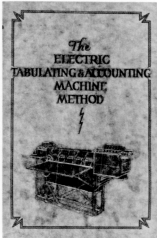

When Watson joined C-T-R, the company's tabulating machine rented for $150 a month. But the rival Powers Company was renting a machine for only $100 a month that was more versatile and automatically printed its results.

Watson challenged two of his engineers to invent a competitive machine. Of the two solutions they offered, the printing mechanism shown here, designed by Clair Lake, was chosen. Watson introduced it from

the stage at a 1919 sales convention. When he threw the switch the machine began printing results, and the salesmen stood on their chairs and cheered.

ONE HUNDRED PERCENT CLUB

INTERNATIONAL BUSINESS MACHINES CORPORATION, ATLANTIC CITY, PERMANENT EXHIBIT.   104590

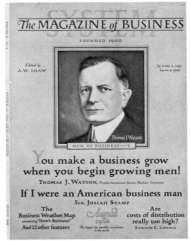

The MAGAZINE of BUSINESS

FOUNDED 1900

Edited by
A.W. SHAW

35 cents a copy
$4 a year

Thomas J. Watson

MEN OF BUSINESS — V.

You make a business grow
when you begin growing men!

THOMAS J. WATSON, President International Business Machines Corporation

If I were an American business man
SIR JOSIAH STAMP

The
Business Weather Map
answering "How's Business"
And 22 other features

August
1926
The largest $4 monthly circulation
in the world

Are
costs of distribution
really too high?
EDMOND E. LINCOLN

The first 100 Percent Club, then known as "Go-Getters," gathered in front of IBM's Atlantic City showroom

Watson believed in salesmanship as a profession—one that he relied on to sustain the company during the Depression. When criticized for hiring more salesmen during those years, he remarked: "Well, you know when a man gets about my age, he always does something foolish. Some men play too much poker, and others bet on horse races, and one thing and another. My hobby is hiring salesmen."

Returning from a trip to Europe aboard the S.S. *Aquitania* in September 1929, Mr. and Mrs. Thomas J. Watson Sr. were photographed with their four children, Jane, Helen, Arthur and Thomas Jr.

Ben Wood and Educational Measurement

"Whatever exists at all, exists in some amount," according to psychologist E. L. Thorndike. His student Ben D. Wood applied this dictum to the measurement of knowledge.

Setting out in 1928 to prepare "an academic inventory of the baccalaureate mind," Wood persuaded Thomas J. Watson to lend him three truckloads of IBM equipment; this became the basis of the Columbia Statistical Bureau, the first university laboratory for educational statistics

Ben D. Wood (at left) with the staff of the Columbia University Statistical Bureau

Benjamin Dekalbe Wood was born in Brownsville, Texas, and had an extensive background in the field of education when he came to Columbia as an instructor in 1921. Interested in more than the statistical aspects of education, Wood also experimented with motion pictures and portable typewriters as teaching aids, and in 1937 authored a plan that became the National Teacher Examination Project.

A tabulating machine was specially modified for Columbia University. The modification enabled the machine to do the difficult mathematical processes necessary for the analysis of educational and psychological tests.

The emitter, a key device, enabled a total from any register to be copied into any of the other nine registers of Columbia's statistical tabulator.

Problems for the "Packard"

Wood pioneered the use of technology in support of educational reform. To help him analyze the results of large-scale educational testing, IBM modified a commercial tabulator so that it could automatically transfer numbers among any of its ten registers. The machine was massive and imposing, and was nicknamed the "Packard."

For other users, Wood's Columbia Statistical Bureau tackled a variety of problems, such as mortality tables and inventories, and for Columbia University's astronomy department, stellar statistics, planetary motions, and the construction of star catalogs.

SUPER COMPUTING MACHINES SHOWN

Columbia Experts Build Device Equal to 100 Mathematicians

ONE DOES 12 PROBLEMS

Another Determines Frequency and One Computes Squares

New statistical machines with the mental power of 100 skilled mathematicians in solving even highly complex algebraic problems were demonstrated yesterday for the first time before a group of psychologists, educational research workers and statisticians in the laboratories of the Columbia University Statistical Bureau in Hamilton Hall.

One of the tabulators exhibited can work out and print the results of as many as twelve difficult problems in just a single rapid operation. It is designed to handle differences and reckon powers of numbers up to the tenth, whereas such machines hiterto have been able to compute only the second power of numbers.

Research Workers Made Them

Richard Warren and Robert M. Mendenhall, research workers at Columbia and statistical consultants for the Carnegie Foundation for the Advancement of Teaching, are responsible for most of the inventions which were first announced at the educator's convention in Atlantic City last week.

These new machines will be a tremendous boon to research. Dr. Ben D. Wood, Director of the Statistical Bureau, said yesterday, through making statistical procedure more accurate, much faster and far less expensive. With the assistance of the new tabu-

New York World, March 1, 1920

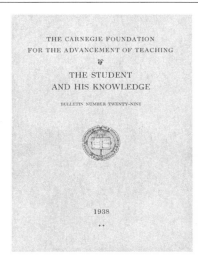

THE CARNEGIE FOUNDATION FOR THE ADVANCEMENT OF TEACHING

THE STUDENT AND HIS KNOWLEDGE

BULLETIN NUMBER TWENTY-NINE

1938

95

Piles of punch cards graphically demonstrate the principle of normal curve distribution. These show the distribution of intelligence test scores of 26,548 Pennsylvania high school seniors; X marks the mean. Conclusions drawn from the tests were published in the book, *The Student and His Knowledge.*

In the early 1930s a Michigan schoolteacher, Reynold B. Johnson, developed the first machine that could sense handwritten pencil marks. Ben Wood saw this invention and convinced Watson that it was the breakthrough necessary to make large-scale testing financially feasible.

IBM hired Johnson (right), and in 1935 the first commercial test scoring machine, the IBM 805, became available.

Test scoring machine dial. The machine registers the score in an analog manner as a weighted comparison of correct and incorrect answers.

Planning the Five-Year Plans

The U.S.S.R. in 1927 embarked on the largest, most comprehensive economic experiment ever undertaken—the Five-Year Plans for total economic control.

Industry and agriculture, transportation, power companies, and government agencies all came under the "Gosplan." Planning at such a scale required an extensive organization for dealing with vast quantities of statistics.

1920   Statistical Machines

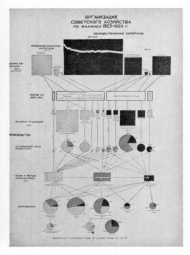

Fourteen branches of the State Bank had tabulating machine installations, making it one of the primary users. This one is at the State Bank in Moscow.

This chart shows the national economy of the U.S.S.R. just prior to the beginning of the Five-Year Plans. At that time, about 70 percent of industry, but only 10 percent of agriculture, was state controlled.

The Statistics of Revolution

The Five-Year Plans implied a level of control that could hardly have been attempted without machines for the rapid assessment of statistics. When Russia started renting tabulating equipment, one of the first installations was at the Central Statistical Bureau.

Other early users were the Soviet Commissariats of Finance, of Inspection and of Foreign Trade, the Grain Trust, the Soviet Railways, Russian Ford, Russian Buick, the Karkov Tractor plant, and the Tula Armament Works. The use of tabulating machinery grew so quickly that in 1929 Russia was reported to be the third largest user, following the United States and Germany.

50 Russians to Learn Tractor

Fifty Russian students will be received into the Special Student Group at the Henry Ford Service School in the near future, according to arrangements made recently.

The students will take a three months' course in Fordson tractor mechanism. They will be selected from worker and student groups and will leave for America some time this month. On their admission to the Special Student Group they will be placed at the Rouge Plant to learn thoroughly the production, assembly, maintenance and repair of the Fordson. When they have completed their course they will return to Russia to serve as instructors in tractor care and repair.

The Fordson tractor line in Detroit

One way Russians learned of punch card machines was through visits to Detroit in the 1920s. Here, Henry Ford poses with a Russian delegation after signing a contract to produce Model A's in Russia.

Posting the results of production and keeping track of the world's largest industrial plan

A Russian-made tractor

Minorsky and Metal Mike

During the First World War, Nicholas Minorsky and others began work on systems that would let a gyrocompass automatically control the steering of a ship, holding it to a prescribed course. In the early 1920s several ships were outfitted with such gyropilots —soon dubbed "Metal Mikes."

In a 1922 paper, "Directional Stability of Automatically Steered Bodies," Minorsky analyzed the problem of control. In its mathematical generality, his paper laid a foundation for another kind of steering—the industrial process control that would come a decade later.

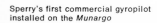

Sperry's first commercial gyropilot installed on the *Munargo*

In 1923 a gyropilot designed according to the principles laid out by Minorsky was installed in the battleship *New Mexico*.

After successful tests, Minorsky concluded that ships could be so well stabilized that their gunfire could be directed from airplanes. He suggested that—since "seeing" would no longer be necessary for steering—the fleet be protected from attack by smoke screens.

In "Directional Stability of Automatically Steered Bodies," Minorsky made a classic contribution to what became the theory of servomechanisms. (The first to apply Taylor's Theorem to a control problem, he was able to solve the differential equation of angular motion to arrive at the best rudder setting at any moment.)

An Augury

[From San Pedro, Cal., to New York, no human hand touched the wheel of the Harry Luckenbach, 5,000 miles. The Gyro-Pilot or "Metal Mike" worked perfectly.—News.]

Oh Metal Mike, whate'er you're like,
Analogies our fancy strike,
An augury we clearly see
To change the course of history.

Your fearless grip, bound not to slip,
Is needed in our statesmanship;
No "still strong man" we know of can
So well tempestuous water span.

A captain new, a wobbling crew,
May matter not as stunts you do;
No blows of fate we shall await
When you are on the Ship of State.

The commonweal is safe we feel,
If you are at the steering wheel;
Oh Metal Mike whate'er you're like,
You'll never, never go on strike!

—J. A.

Homeostasis

Physiologist Walter B. Cannon viewed the animal body as a self-regulating machine. Building on the work done by Claude Bernard in the nineteenth century, Cannon developed the concept of "homeostasis"— the process by which the body maintains itself in a state of internal equilibrium.

Cannon's ideas were well known to Norbert Wiener. In fact, Cannon's *Wisdom of the Body* (1932) may be read as sort of an introduction to Wiener's *Cybernetics* (1948). The essential idea is that homeostatic behavior in animals may be viewed in the same terms as goal-seeking mechanical automata.

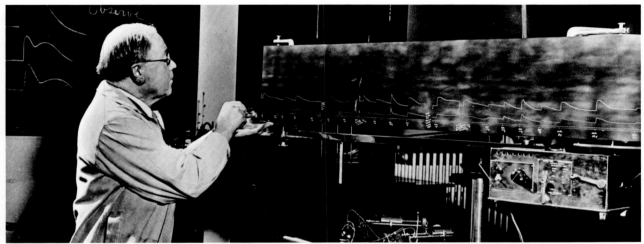

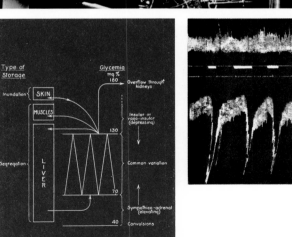

Cannon studying neuromuscular transmission in his laboratory at the Harvard Medical School. The tracings on the kymograph indicate the effects of drugs on the action of nerves.

An example of a feedback process in the human body appears in Cannon's "Organization for Physiological Homeostasis" (1929), in which he coined the term "homeostasis." The diagram shows the regulation of blood glucose levels.

Tracing from May 17, 1911, records "gastric contractions recognized as hunger-pangs."

Cannon photographed with Ivan Pavlov, who established the existence of conditioned reflex through his experiments with dogs.

The English robot Alpha, built about 1932

101

Dark Visions of Machines

By the end of the twenties the world had seen the introduction of mechanized factory operations and the massing of technological force for a world war. Suspicions of machines—of their power, their dominance, their misuse by men—had become a common theme in books, films, and plays.

1930   Logical Automata

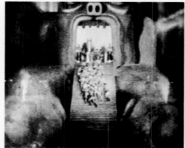

Scene from the Theatre Guild production of Karel Čapek's "R.U.R." (for Rossum's Universal Robots), first performed in Prague in 1921.

In the play, the robot workers manufactured by Mr. Rossum's factory revolt, and kill their makers.

The word *robot* was coined by Karel Čapek's brother Josef from the Czech word *robota,* meaning "forced labor" or "servitude."

Fritz Lang's film "Metropolis," a powerful commentary on the domination of men by machines, was made in 1927. In it, workers held in servitude to the giant industrial machine revolt.

The central dynamo of "Metropolis" is transformed into the god of sacrifice, Moloch.

In his story "The Machine Stops," E. M. Forster describes a vast, comprehensive machine that controls and maintains an entire society—until the machine stops.

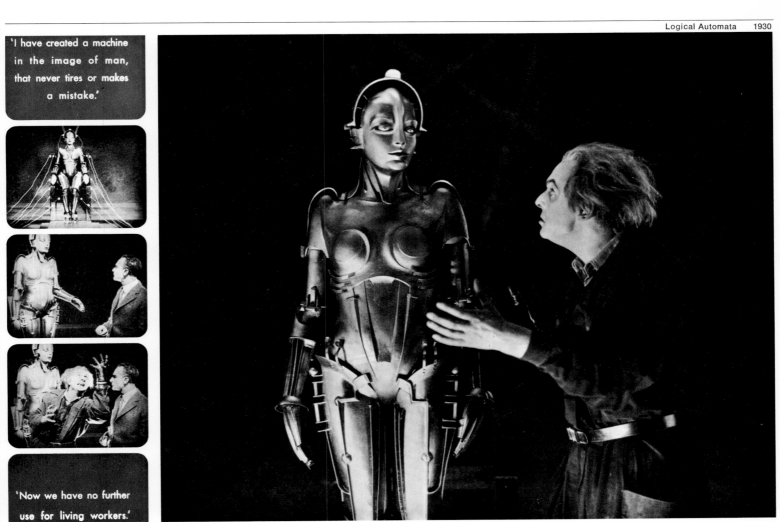

'I have created a machine in the image of man, that never tires or makes a mistake.'

'Now we have no further use for living workers.'

In Metropolis, the scientist, Rotwang, creates a robot worker in the image of the heroine, Maria.

Some Machine Utopias

Visions of a mechanized world flourished in the 1930s. Machines freed mankind from routine labor, or threatened man's humanity by making him a cog in the industrial machine.
The advances in industrial technology made some kind of machine-dominated world seem imminent; but the Depression raised questions about how this world should be run.

A new movement, Technocracy, was widely debated and a popular subject for satires and sermons.

Technocrats advocated that engineers should run the country, abolish the price system, and pay workers in "energy certificates."

1930   Logical Automata

"Utopias now appear much more realizable than one used to think. We are now faced with a very different new worry: How to prevent their realization?"
— Nicolas Berdiaeff

The New York Times Magazine

Section 6

Copyright, 1933, by The New York Times Company.   SUNDAY, JANUARY 8, 1933.   TWENTY PAGES

A CHALLENGE TO TECHNOCRACY

An Examination of the Evidence Which the Technocrats Cite as Proof That a New Economic System Is Needed and a Conclusion That Their Statistics Have Been Highly Inflated

By SIMEON STRUNSKY

THE MACHINE—AND MAN.

The quote above begins Aldous Huxley's *Brave New World*, published in 1932. In this brilliant utopian satire, Huxley draws a picture of the world to come if technology is allowed free rein.

When Charlie Chaplin's "Modern Times" appeared in 1936, the *Herald Tribune* reported: "The mechanized individual goes mad and proceeds to turn the factory into the madhouse that it really always has been."

In both Charlie Chaplin's film, and René Clair's "A Nous la Liberté," the modern assembly line is pictured as turning workers into mechanized automata.

Technocracy first captured the attention of the country when, in 1933, its director, Howard Scott, released an "Energy Survey of North America." Believing that "all social activity must obey the laws of physics," Technocrats urged that the data from this survey be used to run the country.

Archibald MacLeish, writing for *The Nation,* summed up the country's sentiments: "Its statistics are shaky. Its utterances have been half-cocked . . . But for all that, the problem which the word Technocracy unfortunately defines is the vital problem of our time and the hope which the word Technocracy obscures is the first human hope industrialism has offered."

The Technocrats took as their symbol the ancient Chinese generic symbol of Yin-Yang, signifying dynamic balance.

Robots

The robots that appeared in "R.U.R." in 1921 gave their name to a variety of mechanical men built in the 1920s and 30s. In films, or at fairs or festive occasions, they walked and talked and responded to spoken commands. One was human enough to have a dog (naturally, a mechanical one).

Though built in human form, the robots' lifelike qualities were largely an illusion; most of them were demonstrations of remote control by human operators. The first machines that could display initiative—and react, sometimes unexpectedly, to their surroundings—would come in the 1940s.

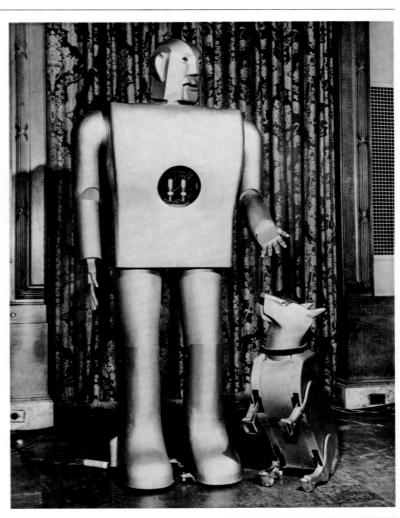

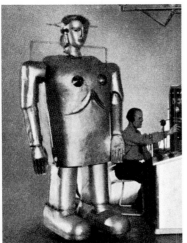

In 1931 Westinghouse built Willie Vocalite to demonstrate remote control techniques. Here, he is enjoying a smoke between duties at the inauguration of air passenger service between New York and San Francisco. He made a speech, wished everyone bon voyage, then helped start the engines.

The Swiss robot Sabor IV and his creator, Auguste Huber, as they appeared in Alfred Chapuis' film "La Féerie des Automates"

The Westinghouse exhibit at the 1939 World's Fair featured Electro and Sparko. Electro could walk, talk, count on his fingers, puff a cigarette, and distinguish between red and green with the aid of a photoelectric cell. His dog, Sparko, would wag its tail, sit up and beg for a hot dog, and bark. (Newsweek assured the public it wouldn't bite.)

Servomechanisms

In the industrial process plants of the 1920s, whether for oil refining or syrup production, operators watched measuring instruments in order to correct the temperature, flow, or pressure.

But in the early 1930s, devices were invented that could use the instruments' output to control the process automatically.

20   *The Literary Digest*   September 9, 1933

A Machine That Bosses Other Machines

ENGINEERS of the Massachusetts Institute of Technology have recently developed a machine "that follows a line like a hound following a trail." Not only can the apparatus—called a "servo-mechanism"—follow an irregular line, but it can actuate other machinery in accordance with the changes of line, and so guide ships, air-planes and complicated mechanical computation machines.

The usefulness of servo-mechanisms is increasing rapidly as they become more accurate and the operations necessary in controlling machinery more complicated. The "mechanical mike" that helped Wiley Post find his way around the world was a servo-mechanism. The term is used to describe any device that guides or controls other apparatus.

A major difference between the operation of Post's automatic pilot and the new M.I.T. apparatus is that the former is designed primarily to follow a straight line, keeping its dependent machinery so adjusted as to pursue a predetermined direct course. Such mechanisms are usually gyroscopically controlled.

The M.I.T. device, which was suggested by Dr. Vannevar Bush and developed under Dr. H. L. Hazen, depends on an entirely different

principle. Instead of gyroscopes it makes use of a light-sensitive cell, so connected that light passing through a narrow slit causes it to deflect a balanced electrical circuit through a small reversible electric motor. The motor in turn drives a worm gear which moves either to the right or left, as necessary, the part of the apparatus carrying the light-sensitive cell.

On one side of the line to be followed the

M.I.T. Official photograph
THE SERVO-MECHANISM AT WORK
Dr. H. L. Hazen of the Massachusetts Institute of Technology watching the device following a curve, part of a complicated engineering problem

paper is darkened; on the other side it is permitted to reflect most of the light falling upon it. The electrical circuit is "balanced" in such a way that when the slit is directly over the edge of the line, covering half dark and half light, no current will flow. When the slit moves toward the dark side, current flows in a direction that causes the motor to pull the machine toward the white. When too much light enters the slit from the white side, the current is reversed and the slit and cell are pushed over toward the dark.

The secret of quick action lies in the motor, which has been specially designed to cut down weight in the armature, the part that rotates. This is so light that the motor, on a tiny current, can reach its full speed of 500 revolutions per minute in one-fiftieth of a second. To prevent the apparatus from running too far and overshooting its mark, it is "damped" by a device that stops the motor quickly when the current ceases to flow.

Among the numerous possible applications of the new device is the automatic steering of ships or aircraft by direct reading of the needle of the magnetic compass. It can also be applied to gun control, to automatic control of industrial processes and many other uses.

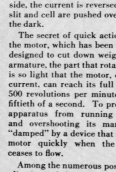

The September 9, 1933, issue of *The Literary Digest* described a light-sensitive servomechanism developed by H. L. Hazen at M.I.T. The machine closely followed a plotted line, and produced a signal that could be used to direct the action of another machine.

Built in 1933, Wolverine-Empire Refining Company's new oil-distilling plant had the latest automatic equipment for industrial process control.

Early photoelectric cell. "Electric eyes" play a part in many process control devices. Photoelectric and selenium cells measure the brownness of coffee beans, the opacity of paper in a paper mill, the shininess of parts on an assembly line.

Hazen and Control

As automatic devices controlled an increasing variety of industrial processes, it was recognized that control could be treated as a topic separate from the process it was controlling. (Leonardo Torres had speculated on this in 1913.)

All control problems, whether the control of a steam turbine, of an airplane, or of a temperature in a chemical process, reduce to the same fundamental principles. H. L. Hazen, who worked on the product integraph with Vannevar

Bush, was the first to treat automatic control systematically in his classic paper, "Theory of Servo-Mechanisms," in 1934.

Harold Black

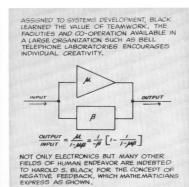

ASSIGNED TO SYSTEMS DEVELOPMENT, BLACK LEARNED THE VALUE OF TEAMWORK. THE FACILITIES AND CO-OPERATION AVAILABLE IN A LARGE ORGANIZATION SUCH AS BELL TELEPHONE LABORATORIES ENCOURAGES INDIVIDUAL CREATIVITY.

$$\frac{OUTPUT}{INPUT} = \frac{\mu}{1-\mu\beta} = \frac{1}{-\beta}\left[1 - \frac{1}{1-\mu\beta}\right]$$

NOT ONLY ELECTRONICS BUT MANY OTHER FIELDS OF HUMAN ENDEAVOR ARE INDEBTED TO HAROLD S. BLACK FOR THE CONCEPT OF NEGATIVE FEEDBACK, WHICH MATHEMATICIANS EXPRESS AS SHOWN.

Taylor automatic controller. When a change occurred in temperature or pressure, flow or liquid level, this device automatically produced a counterresponse calculated to maintain a state of internal balance.

Lent by George Howard

Harold Black spent years trying to produce distortion-free amplification in telephone transmission. Then he thought of reversing some of the output of an amplifier and feeding it back into the input. He found that this arrangement removed distortion. Shown is Black's diagram and mathematical equation for this extremely important idea in control: electronic negative feedback.

Black feedback amplifier. Embodying the principle of negative feedback, these amplifiers were first used in 1936 as repeaters in the coaxial cables between New York and Philadelphia. They made it possible to carry hundreds of telephone messages simultaneously with great clarity.

Taylor automatic controller.

Social Security

One of Franklin D. Roosevelt's presidential campaign promises in 1932 was ''social justice through social action.'' One of the major instruments of that social action was the Social Security Act of 1935.

The Act made it necessary for the government to maintain the employment records of twenty-six million people. Not since World War I had personal data been organized on such a scale.

1930   Statistical Machines

President Franklin D. Roosevelt signing the Social Security Act

Clerks punch information received from employers into cards.

Lining up for Social Security information, 1936

Accounting for Everybody

The "world's biggest bookkeeping job" was done in a Baltimore brick loft building, chosen because it had 120,000 square feet of floor space, and was structurally strong enough to bear the weight of 415 punching and accounting machines.

A production line was set up to punch, sort, check, and file half a million cards a day. The collator, developed by IBM especially for the job, became a widely used device in government and business generally.

The machines established the ability of the government to implement national programs in individual terms, for example, the introduction of the withholding tax in 1943.

SUNDAY NEWS, JANUARY 10, 1937

BIGGEST BOOKKEEPING JOB BEGINS

Social Security Board Has Gigantic Task

By GUY RICHARDS.

IBM 077 collator, developed for the Social Security program. H. J. McDonald, who sold the account, remembers IBM President Watson

CHOOSEY

SORTER

Boris Artzybasheff's impressions of a sorter and collator appeared in a booklet explaining the subscription service of Time Inc.

The IBM 077 collator was developed for the Social Security program. H. J. McDonald, who sold the account, recalls that IBM President Watson ordered the development of the collator because "The Social Security agency punched cards from records sent in by employers all over the country. There were millions and millions of them, and if we hadn't had some way of putting them together we would have been lost; we just couldn't have done it."

Mrs. Ida Fuller, of Ludlow, Vermont, receives the nation's first Social Security benefit check in 1937.

Hooton: *The American Criminal*

A physical anthropologist at Harvard, Earnest A. Hooton, was interested in the effect man's anatomy has on his behavior. Starting in 1927, he and his investigators took measurements in the prisons of ten states to study the relation between a criminal's physical makeup and the nature of his crime.

Hooton collected social, racial, and psychological data, facial characteristics and some twenty physical measurements. The resulting mass of data was analyzed on punch card tabulating machines.

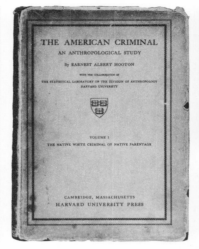

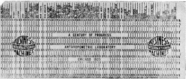

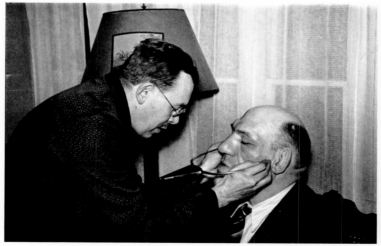

Recording sheet used in Hooton's study of criminals. Hooton, recalling the stacks of information coming in from his investigators, said: "there is only one kind of ostrich capable of digesting such a mass of data. Its name is Hollerith and it may be described as a species of electric sorting, counting, and tabulating machine which swallows cards with holes punched in them and spews up figures and statistical tables. It is something more than human, but rather less than divine."

Hooton measuring the French wrestler known as "The Angel"

In 1933 Hooton published his findings in an exhaustive report, *The American Criminal*.

Born in Clemansville, Wisconsin, Hooton studied anthropology at the University of Wisconsin and at Oxford. His publications include such titles as *Why Men Behave Like Apes and Vice-Versa* and *Young Man, You Are Normal*.

Hooton and C. W. Dupertuis set up an Anthropometric Laboratory at the 1933 Century of Progress Exposition in Chicago to demonstrate anthropometric techniques and the statistical analysis of data. Their goals were very much like those of Sir Francis Galton at the International Health Exhibition in London fifty years earlier.

America Speaks

The measurement of opinion was attempted first in market research, then in political prediction. In the famous 1936 battle for election polling supremacy, George Gallup, canvassing only 30,000 people, defeated the statisticians of the *Literary Digest,* who polled more than 2,300,000.

His weapon was a new set of tools: techniques for identifying and polling a small but statistically valid cross section of the population.

George Horace Gallup studied journalism at the State University of Iowa. After holding a series of academic posts he joined the advertising firm of Young and Rubicam, and in 1935 founded the American Institute of Public Opinion.

Gallup checking the information on a completed questionnaire. The data is punched in a card, and the cords sorted by machine.

This table, worked out by Harvard Professor Theodore Brown, gives the size of sample necessary for accuracy in random sampling.

Brown's table shows, for example, that (997 times out 1000) a prediction based on only nine hundred people will be accurate within 5 percent.

Leontief and Input-Output

Given any economic event—for example, the passage of a housing bill by Congress—how can you foretell the effect it will have on the nation's economy?

Wassily W. Leontief devised a mathematical way to predict the consequences of economic decisions. But to do the vast amount of calculations necessary, he would need more powerful kinds of computing machines.

Table 5. – Quantitative Input and Output Relations in the Economic System of the United States, 1919
(Unit: million dollars)

[Large input-output data table: "DISTRIBUTION OF OUTLAYS (INPUT) OF CLASSES LISTED AT TOP OF TABLE" (rows) versus "DISTRIBUTION OF OUTPUT OF CLASSES LISTED AT LEFT OF TABLE" (columns), numbered 1 through 46, listing industries: Agriculture; Flour and grist mill products; Canning and preserving; Bread and bakery products; Sugar, glucose, and starch; Liquors and beverages; Tobacco manufactures; Slaughtering and meat packing; Butter, cheese, etc.; Other food industries; Iron mining; Blast furnaces; Steel works and rolling mills; Other iron and steel and electric mfrs.; Automobiles; Non-ferrous metal mining; Smelting and refining; Brass, bronze, copper, etc., mfrs.; Non-metal minerals; Petroleum and natural gas; Refined petroleum; Coal; Coke; Manufactured gas; Electric utilities; Chemicals; Lumber and timber products; Other wood products; Paper and wood pulp; Other paper products; Printing and publishing; Yarn and cloth; Clothing; Other textile products; Leather tanning; Leather shoes; Other leather products; Rubber manufactures; Industries n.e.s.; Construction; Transportation (steam railroads); Imports; Wages and salaries; Capital and entrepreneurial services; Total services; Undistributed: Taxes, Other, Total; Gross total outlays; Net total outlays.]

Harvard economist Wassily W. Leontief

Leontief's first input-output chart. "It is typical of the Roosevelt government," Leontief recalls, "that when they came to me in 1941—before the United States entered the War—they had already decided to make use of the input-output approach in a study of the implication of eventual disarmament in the United States *after* the War."

Leontief's typescript for his paper, "Interrelation of Prices, Output, Savings, and Investment," published in *The Review of Economic Statistics* in 1937.

It shows two of the three sets of equations that made up his proposed input-output scheme.

Wassily Leontief, born in Russia in 1906, was educated at the universities of Leningrad and Berlin before joining the faculty of Harvard University in 1931.

An Economic Model

The state of a country's economy is usually judged by its end products—so many houses, so many automobiles, so many telephones. But most industries don't make end products; they provide materials and services to industries that do make end products.

To get a statistical picture of the economy, Leontief made tables which showed the cost of the products that each industry buys from—and sells to—every other industry. Through calculation, he could then trace the end result of any economic action.

The more kinds of industries the input-output tables include, the more calculation is required. Larger and larger grids have been attempted as the power of computing machinery has grown.

John Wilbur at his Equation Solver.

Plate from John Wilbur's Simultaneous Equation Solver.

Leontief began with a 42-sector input-output model, but the calculation required about 30 million multiplications, which exceeded the capabilities of any system of calculation, hand or mechanical, available in the mid-thirties.

So Leontief simplified the data into a 10-sector grid, but even that would have required 450,000 multiplications, or as he reckoned, two years at 120 multiplications per hour. Instead, he used the Wilbur machine at M.I.T.

Made of tilting steel plates (representing unknowns) and steel tapes (representing equations to be solved), the machine could solve nine simultaneous equations. Leontief recalls: "You could really change the coefficients slightly by simply sitting on the frames, and if they did not give too much this meant that the solution was relatively stable."

Eckert's "Mechanical Programmer"

In 1933 Wallace J. Eckert felt the "time was now ripe for the establishment of a scientific laboratory of a revolutionary nature." Knowing of Thomas Watson's interest, he submitted to IBM a list of essential equipment. Within a few weeks, the machines were delivered to the attic of Columbia University's Pupin Hall, later known as the Astronomical Computing Bureau.

Eckert linked together, for the first time, different kinds of punch card accounting machines to allow complex calculations.

He recalled: "About this time the 601 multiplying punch, the credit balance accounting machine and the summary punch appeared in the IBM line—in other words, mechanical reading, writing, and arithmetic were at hand."

1930   Calculating Machines

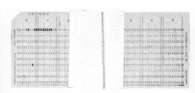

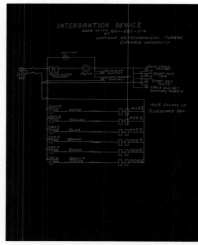

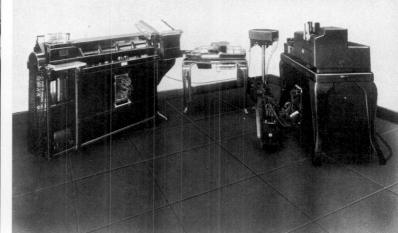

Eckert, a young assistant in astronomy, requisitioned the first calculator (an electric Monroe) to be used in a science class at Columbia. "When I started in 1926 as an assistant at Columbia," he recalled, "the logarithm was just being put to bed and the desk calculator was beginning to take over computation work. It was an exciting time as we began to see the real possibilities of automatic computation emerging."

In 1931 Eckert began using Columbia's statistical tabulator on weekends for his astronomical calculations.

Eckert frequently returned to his chief astronomical interest: lunar theory. In 1948 he predicted the positions of the moon using the theory which was the basis for E. W. Brown's classic tables. The results were so accurate that previously undetected irregularities in the earth's rotation were discovered.

The first installation at the Columbia Department of Astronomy showing the 601 multiplying punch, the credit balance accounting machine, and the summary punch interconnected by Eckert's "mechanical programmer."

The Calculation Control Switch

To run these machines as a unit, Eckert made a "mechanical programmer." It controlled the pluggable relay box taken from Columbia's recently dismantled statistical tabulator.
In devising this scheme, Eckert took an important step into the zone that separated the calculator of those days from the concept of the computer. His unique arrangement of machines became the nucleus of scientific computing at Columbia University.

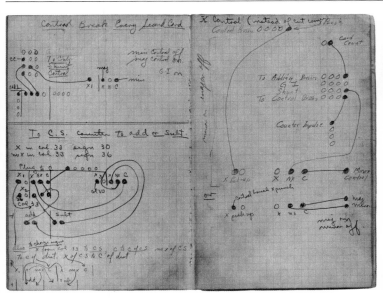

Eckert's notebook of "machine ideas" open to some plugging diagrams for his control switch

Eckert described his "mechanical programmer": "There were about twenty successive steps or settings for the switch and we filed nicks in the disks to tell the different machines what to be doing at any time."

In 1940 Eckert left Columbia to become director of the Nautical Almanac Office. With war approaching, the immediate job was to produce quickly an Air Almanac for the Army's pilots. Eckert set up punch card machines to do the calculation and print out the results in a form that could be directly photo-offset.

Astronomical Computing Bureau. In 1943 the facilities were greatly enlarged and calculations were undertaken for all aspects of the war effort, including evaluations of the B-29 fire control system and computations on nuclear fission for the Manhattan Project.

The Bush Differential Analyzer

In the late 1920s Vannevar Bush was trying to solve equations associated with power failures. To handle such "differential" equations, he built, in 1930, the first automatic computer that was general enough to solve a wide variety of problems.

Bush called it a "differential analyzer," and helped build similar machines at Aberdeen and at the Moore School in Philadelphia. They were accurate to within 0.05 percent.

1930   Calculating Machines

116

Vannevar Bush with his first differential analyzer. "I was trying to solve some of the problems of electric circuitry, such as the one connected with failures and blackouts in power networks," he reported, "and I was thoroughly stuck because I could not solve the tough equations the investigation led to."

Bush attended Tufts College, the Massachusetts Institute of Technology, and Harvard University. In 1919 he joined the faculty of M.I.T., where he developed his product integraph and differential analyzer. From 1932 to 1938 he served as the Dean of the School of Engineering and vice-president of the Institute. As director of the Office of Scientific Research and Development during World War II, Bush played a major role in guiding the United States war effort.

Analog Computing

Before Bush's machine, analog computers solved only particular kinds of problems. The "network analyzers" built by the power companies in the 1920s were computing simulators—in effect, scale models of a power network.

But Bush's differential analyzer was the first general equation solver. It was very successful. And to many people it began to appear that such big, general-purpose analog computers would dominate scientific calculation in the future.

In 1935 Bush began to build an even more general machine using electrical components.

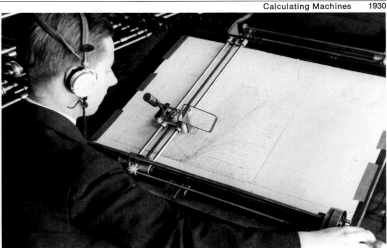

A.C. network analyzer designed in 1929 and installed at Westinghouse in 1930. Built by power companies to analyze load requirements, there were fifty in operation in the middle 1950s.

Predicting the result of large surges in order to avoid blackouts is a more difficult problem, however, and the need for these answers led Vannevar Bush to the invention of his differential analyzer.

The Bush differential analyzer originally gave its solutions in the form of curves; later it was adapted to have a five-register numerical output (seen in foreground).

After the differential equation is set up in the machine, operators stationed at plotting tables enter data by keeping a moving pointer on a curve as the solution proceeds.

In Bush's differential analyzer #2, punched paper tapes set up the connections to be made, the values of gear ratios, and values of initial conditions.

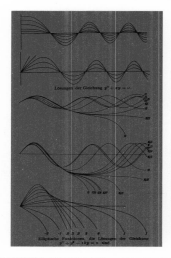

118

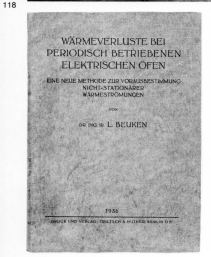

D.C. calculating board from the 1920s. It was one of the earliest computing simulators designed to solve problems in power transmission.

Heat flow problems arise in furnace design as well as in many other areas of industry. In his 1936 thesis, "Heat Loss in Periodically Activated Electric Ovens," and in the model he built to prove his ideas, L. Beuken showed that electricity could be used to simulate heat flow. Large machines based on this principle are still in use.

Solutions of equations drawn by a differential analyzer built in 1938 by Svein Rosseland in Oslo. It was one of the many machines based on Bush's basic design.

The building of the differential analyzer depended on the invention of the "torque amplifier" by H. W. Nieman of the Bethlehem Steel Company. In the photo, the weak force exerted by the lady is amplified and becomes a powerful rotation at the other end.

Meccano and Quantum Mechanics

British physicist Douglas R. Hartree was the first to use an automatic calculator for problems in atomic theory. During the summer of 1933, he went to M.I.T. and used Vannevar Bush's differential analyzer for the first stages of the computa-tions for "Approximate Wave Functions for Mercury." The work was a success, and on his return to England, he undertook with Arthur Porter the building of a model of the machine.

The model was originally built from about £20 worth of Meccano parts to demonstrate the principles of Bush's differential analyzer. Nevertheless, it was so successful (the accuracy was the same as Bush's product integraph—within about 2 percent) that it was used to find approximate solutions of some problems in wave mechanics. Based on the success of the model, Hartree directed the construction of a full-scale differential analyzer the following year at the University of Manchester.

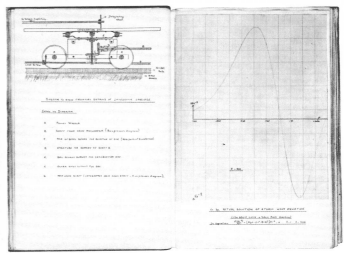

Douglas R. Hartree and Arthur Porter with their differential analyzer. Hartree commented on how the idea came about: "My first impression on seeing the photographs of Dr. Bush's machine was that . . . someone had been enjoying himself with an extra large Meccano set."

Arthur Porter's master's thesis: "The Construction of a Model Mechanical Device for the Solution of Differential Equations, with Applications to the Determination of Atomic Wave Functions" (1934).

Zuse

In 1936 Konrad Zuse started to build a calculator in the living room of his parents' apartment in Berlin. He called his machine the Z1. By 1941 he had completed two more relay calculators, the Z2 and Z3.

With no knowledge of what was being done in Britain and the United States, Zuse pioneered some of the basic ideas of automatic computing. Binary arithmetic, floating decimal point, and program control by punched tape were incorporated in these early machines.

These machines were all destroyed in the war, but Zuse managed to save a new calculator he had partially completed—the Z4. Zuse went on to design improved models and made them the basis of a successful computer manufacturing company, Zuse KG.

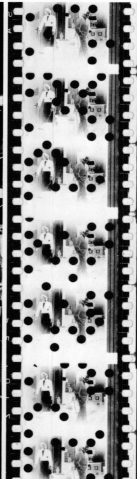

The Z1 in Zuse's parents' living room

Zuse with the program reader of his Z4. Instructions for the machines were punched into discarded 35mm movie film.

An engineer by training, Konrad Zuse developed in 1945-46 the "plan calculus," a forerunner of modern programming languages.

The Switch to Base Two

In 1937 Claude E. Shannon, for his master's thesis at M.I.T., described a way of using symbolic logic to improve electrical switching circuits. In one example, he showed how to simplify an "Electric Adder to the Base Two." That same year, George Stibitz built such an adder on his kitchen table with "some relays from a scrap pile at Bell Labs where I then worked." He named it after the kitchen table: The Model "K." Neither man knew of the other's work.

During the next decade, Stibitz built relay calculators of increasing complexity. Shannon's paper, as it turned out, proved that programming an electronic digital computer would be a problem, not in arithmetic, but in logic.

Principia Mathematica

In 1910 *Principia Mathematica* by Bertrand Russell and Albert North Whitehead presented the idea that logic is the foundation of all mathematics. It develops the calculus of propositions, solving problems in terms of statements that are either true or false.

Claude E. Shannon

ELECTRIC ADDER TO THE BASE TWO

A circuit is to be designed that will automatically add two numbers, using only relays and switches. Although any numbering base could be used the circuit is greatly simplified by using the scale of two.

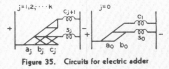

Figure 35. Circuits for electric adder

Figure 36. Simplification of figure 35

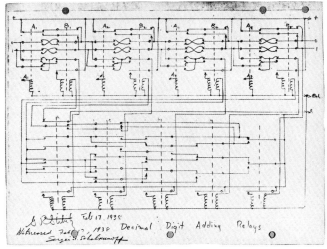

Shannon's thesis, published in the *Transactions of the American Institute of Electrical Engineers,* provided a theoretical basis for the entire set of operations that would be designed into electronic digital computers.

While a graduate student, Shannon took a part-time job operating a Bush differential analyzer. The analyzer's relay circuits needed frequent attention, and Vannevar Bush suggested to Shannon that the design of such circuits would make a good thesis subject.

Relays and crossbar switches connect two telephones in direct dialing. The first American crossbar central office opened in Brooklyn on February 13, 1938. At just about the same time, George Stibitz was beginning work on his "complex number calculator" (Model I), which consisted largely of relays and crossbar switches.

In his Model I relay calculator, Stibitz used a "mixed binary decimal" notation. This diagram is for a circuit that will add together two decimal digits.

The methods of symbolic "true or false" logic described by Russell and Whitehead were, in 1937, shown to have practical application to the design of electrical circuits (in Shannon's thesis, "Symbolic Analysis of Relay and Switching Circuits").

Aiken and the A.S.C.C.

Harvard graduate student Howard H. Aiken proposed, in 1937, that a new kind of calculating machine be built. He wrote: "there exist problems beyond our ability to solve, not because of theoretical difficulties, but because of insufficient means of mechanical computation."

Aiken was thinking of linking together Monroe calculators on the "player piano" principle to create a large scientific calculator, but Harvard astronomer Harlow Shapley sent him to IBM.

1930   Calculating Machines

In his 1937 proposal, Aiken listed some of the fields that needed more powerful calculating methods: vacuum tube design, wave mechanics, physics of the upper atmosphere, astronomy, relativity, and the "science of mathematical economy." When completed, the calculator was mainly used by the U.S. Navy for ballistics and ship design, but it also solved some problems in lens design, and did work for Wright Patterson Air Force Base and for the Atomic Energy Commission.

The co-inventors of the Harvard Calculator (from left): Francis E. Hamilton and Clair D. Lake of IBM, Howard H. Aiken of Harvard, and Benjamin M. Durfee of IBM.

Mark 1

In 1938 IBM president Thomas J. Watson Sr., impressed with Aiken's ideas, assigned engineer Clair D. Lake to supervise the designing and building of a large machine for Harvard University. It was completed in 1944. Named the "Automatic Sequence Controlled Calculator," it became known at Harvard as the "Mark I." It was the first automatic, general-purpose, digital calculator.

Mark I became the basis for the Harvard Computation Laboratory. There, Aiken encouraged people in a variety of fields—among them economics, insurance, physics, and linguistics—to use the machine in solving their problems.

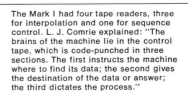

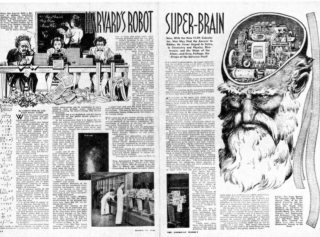

After running test problems at IBM's Endicott plant for a year, the Mark I was delivered to Harvard in February 1944. By April, it was in full-scale operation tackling ballistics problems for the Navy.

Numerical data could be introduced into the machine via punched paper tape, punch cards or manually set dial switches.

The Mark I had four tape readers, three for interpolation and one for sequence control. L. J. Comrie explained: "The brains of the machine lie in the control tape, which is code-punched in three sections. The first instructs the machine where find its data; the second gives the destination of the data or answer; the third dictates the process."

Some tapes contained the instructions for the particular problem being run;

others held standard subroutines which could be stored and used when a problem required them.

The Universal Turing Machine

Alan M. Turing had just graduated from Cambridge when he published a crucial theorem in mathematical logic in terms of an idealized computing machine.

In his most famous contribution to mathematics, "On Computable Numbers," Turing offered a proof that, even when given a "fixed and definite process" for solving a set of problems, some of those problems still cannot be solved. This was contrary to the prevailing view of famous mathematician David Hilbert.

1930   Logical Automata

Alan M. Turing

124

appears that the men who I wrote to, and whom I asked to communicate the paper since had gone to China, and moreover the letter seems to have been lost in the post, since a second letter reached his daughter. Meanwhile a paper has appeared in America, written by Alonzo Church, doing the same things in a different way. Mr. Newman and I have decided however that the method is sufficiently different to warrant the publication of my paper too. Alonzo Church lives at Princeton so I have decided quite definitely about going there.

[No. 40]

KING'S COLLEGE,
CAMBRIDGE.

29 May
[1936]

Dear Mother,
By all means let Nancy have her holiday at the time you suggest. I don't think I shall be wanting to stay in Guildford much just then, and anyway we should be able to manage perfectly well without her.

I have just got the main paper ready & sent in. I imagine it will appear in October or November. The situation with regard to the note for Comptes Rendus was not so good. Y*

[On Computable Numbers. G.B.]

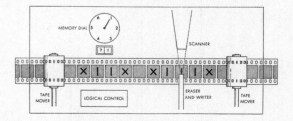

Diagram of a Universal Turing Machine
from *Scientific American,* April 1955

In a 1936 letter to his mother, Turing wrote that his paper "On Computable Numbers" would be published that fall, as it was significantly different from one on the same subject that had just been published by Alonzo Church. Since Church taught at Princeton, Turing reported, "I have decided quite definitely about going there."

Once at Princeton, Turing wrote home: "The mathematics department here comes fully up to expectations. There is a great number of the most distinguished mathematicians here, J. v. Neumann, Weyl, Courant, Hardy, Einstein, Lefschetz, as well as hosts of smaller fry."

Alan Mathison Turing was born in London in 1912. At the age of twenty-six he turned down a job as von Neumann's assistant to return to England, where he joined the British Foreign Office. From 1948 until his death at the age of forty-two, Turing was on the faculty of Manchester University.

"On Computable Numbers"

Turing defined a "definite process" as "something that could be done by an automatic machine," and went on to describe in theory a machine which could do any calculation that could be done by a human. This "Universal Turing Machine" contained ideas later incorporated into all general computing machines.

Given a suitable series of its four simple instructions, Turing's machine would imitate any other computing machine; in effect, it provides a standard for measuring the complexity of any computer.

His tutor, M. H. A. Newman, wrote in 1955: "It is difficult today to realize how bold an innovation it was to introduce talk about paper tapes and patterns punched in them into discussions of the foundations of mathematics."

3 MILES RACE WON BY ONE FOOT
FAST TIMES AT WALTON
By A Special Correspondent

Excellent weather and a track in first-class condition considerably helped athletes to record fast times in the open handicap meeting at Stompond-lane sports ground, Walton, yesterday.

Receiving the extraordinary start of 20 yards, K. Owen, of the local club, had no difficulty in winning the 120 yards handicap in 12.1sec from L. King and L. C. Lewis, the R.A.F. 440 yards champion, who were second and third.

The Three Miles was a thrilling race in which C. G. Scott (Surrey A.C.), who started from the 10-yards mark, beat the scratch man, A. M. Turing (Walton A.C.), by one foot in the last stride in 15min 51sec.

---

230   A. M. TURING   [Nov. 12,

ON COMPUTABLE NUMBERS, WITH AN APPLICATION TO THE ENTSCHEIDUNGSPROBLEM

By A. M. TURING.

[Received 28 May, 1936.—Read 12 November, 1936.]

The "computable" numbers may be described briefly as the real numbers whose expressions as a decimal are calculable by finite means. Although the subject of this paper is ostensibly the computable *numbers*, it is almost equally easy to define and investigate computable functions of an integral variable or a real or computable variable, computable predicates, and so forth. The fundamental problems involved are, however, the same in each case, and I have chosen the computable numbers for explicit treatment as involving the least cumbrous technique. I hope shortly to give an account of the relations of the computable numbers, functions, and so forth to one another. This will include a development of the theory of functions of a real variable expressed in terms of computable numbers. According to my definition, a number is computable if its decimal can be written down by a machine.

In §§ 9, 10 I give some arguments with the intention of showing that the computable numbers include all numbers which could naturally be regarded as computable. In particular, I show that certain large classes of numbers are computable. They include, for instance, the real parts of all algebraic numbers, the real parts of the zeros of the Bessel functions, the numbers π, e, etc. The computable numbers do not, however, include all definable numbers, and an example is given of a definable number which is not computable.

Although the class of computable numbers is so great, and in many ways similar to the class of real numbers, it is nevertheless enumerable. In § 8 I examine certain arguments which would seem to prove the contrary. By the correct application of one of these arguments, conclusions are reached which are superficially similar to those of Gödel†. These results

† Gödel, "Über formal unentscheidbare Sätze der Principia Mathematica und verwandter Systeme, I", *Monatshefte Math. Phys.*, 38 (1931), 173–198.

THE JOURNAL OF SYMBOLIC LOGIC
Volume 1, Number 3, September 1936

FINITE COMBINATORY PROCESSES—FORMULATION 1

EMIL L. POST

The present formulation should prove significant in the development of symbolic logic along the lines of Gödel's theorem on the incompleteness of symbolic logics[1] and Church's results concerning absolutely unsolvable problems.[2]

We have in mind a *general problem* consisting of a class of *specific problems*. A solution of the general problem will then be one which furnishes an answer to each specific problem.

In the following formulation of such a solution two concepts are involved: that of a *symbol space* in which the work leading from problem to answer is to be carried out,[3] and a fixed unalterable *set of directions* which will both direct operations in the symbol space and determine the order in which those directions are to be applied.

In the present formulation the symbol space is to consist of a two way infinite sequence of spaces or boxes, i.e., ordinally similar to the series of integers $\cdots$, $-3, -2, -1, 0, 1, 2, 3, \cdots$. The problem solver or worker is to move and work in this symbol space, being capable of being in, and operating in but one box at a time. And apart from the presence of the worker, a box is to admit of but two possible conditions, i.e., being empty or unmarked, and having a single mark in it, say a vertical stroke.

One box is to be singled out and called the starting point. We now further assume that a specific problem is to be given in symbolic form by a finite number of boxes being marked with a stroke. Likewise the answer is to be given in symbolic form by such a configuration of marked boxes. To be specific, the answer is to be the configuration of marked boxes left at the conclusion of the solving process.

The worker is assumed to be capable of performing the following primitive acts:[4]

(a) *Marking the box he is in* (assumed empty),
(b) *Erasing the mark in the box he is in* (assumed marked),
(c) *Moving to the box on his right*,
(d) *Moving to the box on his left*,
(e) *Determining whether the box he is in, is or is not marked*.

[3] The set of directions which, be it noted, is the same for all specific problems and thus corresponds to the general problem, is to be of the following form. It is to be headed:

Start at the starting point and follow direction 1.

Received October 7, 1936. The reader should compare an article by A. M. Turing, *On computable numbers*, shortly forthcoming in the *Proceedings of the London Mathematical Society*. The present article, however, although bearing a later date, was written entirely independently of Turing's. *Editor*.

[1] Kurt Gödel, *Über formal unentscheidbare Sätze der Principia Mathematica und verwandter Systeme I*, *Monatshefte für Mathematik und Physik*, vol. 38 (1931), pp. 173–198.
[2] Alonzo Church, *An unsolvable problem of elementary number theory*, *American Journal of Mathematics*, vol. 58 (1936), pp. 345–363.
[3] Symbol space, and time.
[4] As well as otherwise following the directions described below.

103

"On Computable Numbers" by Alan Turing, and "Finite Combinatory Process—Formulation 1," by American Emil Post were both done independently in 1936. Post suggests a computation scheme by which a "worker" can solve all problems in symbolic logic by performing only machinelike "primitive acts." Remarkably, the instructions given to the "worker" in Post's paper and to a Universal Turing Machine were identical.

America's answer!

PRODUCTION

JEAN CARLU

*U.S. GOVERNMENT PRINTING OFFICE 1941—O—454542

DIVISION OF INFORMATION
OFFICE FOR EMERGENCY MANAGEMENT
WASHINGTON, D.C.

127

Self-Regulating Systems

Wartime requirements accelerated the development of mechanisms for prediction and control. Hardware was built that measured the difference between actual performance and the desired result—then "homed-in" on the target.

With the development of such goal-seeking machines, the idea of purposeful behavior could be defined in engineering terms.

1940   Logical Automata

128

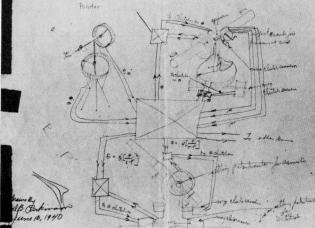

David B. Parkinson's original drawing for the M-9, made on June 18, 1940

Shortly after Dunkirk, Parkinson dreamt that he was a member of a Dutch antiaircraft battery which possessed a marvelous robot mechanism. This machine tracked the German Stukas, computed the information, and then directed the aiming and firing of antiaircraft guns with uncanny accuracy.

Even awake the idea seemed feasible, and Parkinson worked with his colleagues at Bell Telephone Laboratories—Clarence A. Lovell and Bruce T. Weber—to develop it in reality: the M-9 gun director.

Designed in 1942, the M-9 was an awesome calculating robot that played a large part in the defense of London against V-1 "Flying Bombs."

After radar locked-in on the target, the system regulated itself, calculating ballistic curves and directing 90mm guns with great accuracy. In the October 1944 Allied defense of Antwerp, 4883 V-1's were engaged, and 4672 were shot down—an accuracy of 95.7 percent. Out of 91 V-1 attacks on the last day, 89 were destroyed!

The M-9 epitomized the goal-seeking machine. It brought together automatic calculation using negative feedback with electronically controlled servomechanisms.

"Operational amplifier" of the type used in the M-9. Introduced around 1940, it allowed mathematical operations to be performed with voltages as the variables. Its invention made the electronic analog computer practicable, and precluded the need for further development of such mechanical machines as the Bush differential analyzers.

Behavior, Purpose, and Teleology

Self-regulating robots, such as antiaircraft gun directors, brought together the technologies of gyroscopic guidance, servomechanisms, and analog computation. Each of these had been used for some industrial purpose before World War II; however, their combination introduced a sophistication that was to produce new techniques in the control of automatic factories.

Many mathematicians had been intimately involved in these wartime developments. Their speculations on the future goals of an engineering-oriented society were greeted with a new level of public interest.

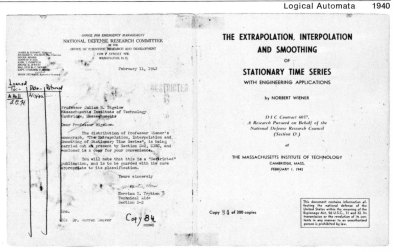

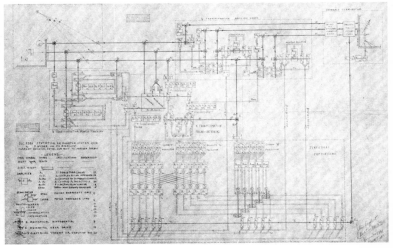

The World War II Norden bombsight was a carefully balanced combination of analog computation and guidance control.

It had two parts: a stabilizer that controlled the plane during the ten to thirty seconds of the bombing run, and the bombsight proper that contained a telescope and computer. Both parts contained gyroscopes.

The intense activity related to gun directors and computers during World War II resulted in a joint effort by Howard Aiken, John von Neumann, and Norbert Wiener to found a Teleological Society devoted "to the study of how purpose is realized in human and animal conduct and . . . how purpose can be imitated by mechanical and electrical means."

Diagram for Norbert Wiener's and Julian Bigelow's statistical antiaircraft gun director system, November 5, 1941

M.I.T. mathematician Norbert Wiener approached gun directing as a statistical problem. "Bigelow and I felt that we could safely go ahead with the treatment of the human links in the control chain as if they were pieces of feedback apparatus."

Although the designs were not perfected, the theoretical basis of the work was circulated in this classified document: *The Extrapolation, Interpolation and Smoothing of Stationary Time Series* (which became known—after the color of its cover—as the "yellow peril").

Wiener wrote that his aim was to bring together two separate fields: time series in statistics and communication engineering. This statistical approach considerably influenced Claude Shannon in his development of information theory, published in 1948.

Aircraft Simulators

Simulators had been built as flight trainers to give pilots realistic responses to their manipulation of the controls. But with the advanced control technologies of World War II came the idea of building a more elaborate aircraft simulator to test the aircraft's design even before it was put into production.

In 1944 the U.S. Navy asked Gordon Brown to build a simulator for multi-engine aircraft in his unique M.I.T. Servomechanisms Laboratory.

For "real time" simulation, an electromechanical analog proved too slow. Perry Crawford, who in his M.I.T. master's thesis had shown how electronic digital computing techniques could be used for automatic fire control, suggested the same approach for simulation to Jay Forrester, the project's leader. He developed the idea, and by 1946 the contract to develop an aircraft simulator had evolved into a project to design a digital computer, Whirlwind I.

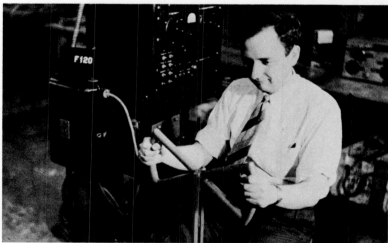

Link Aviation employees gather for a portrait with one of the first flight trainers.

Trial model of the multiengine flight simulator at the M.I.T. Servomechanisms Laboratory

Project Pigeon

During the war, behavioral psychologist B. F. Skinner demonstrated an automatic homing system which would guide a bomb directly to its target.

Skinner's control system used a lens in the nose of the bomb to throw an image of the approaching target on a ground-glass screen. Inside, a pigeon trained to recognize the desired target pecked at it with its beak. If the target's image moved off center, the pigeon's pecking tilted the screen, which moved the bomb's tail surfaces, which corrected the bomb's course. To improve accuracy, Skinner used three pigeons to control the bomb's direction by majority rule. According to him, the system was resistant to jamming, simply built, and needed no materials in short supply.

Despite these advantages, the military review board would not let the idea get off the ground.

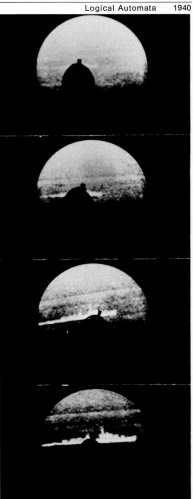

Demonstration model of the three-pigeon guidance system

Pigeons jacketed for testing

Motion picture frames of a pigeon making a simulated bombing approach toward a ship at six hundred miles per hour

ENIAC At The Moore School

The first general-purpose electronic calculator was formally dedicated at the Moore School of Electrical Engineering at the University of Pennsylvania in February 1946. Built to do ballistic calculations for the U.S. Army, it was named the "Electronic Numerical Integrator and Computer"— and called, after its initials, the ENIAC.

The ENIAC in operation at the Moore School prior to its installation at Aberdeen. In the foreground are co-inventors J. Presper Eckert Jr. (left) and John W. Mauchly.

The ENIAC's main job (and the one it was best suited for) was the lengthy and repetitive calculation of ballistic tables. However, the ENIAC was used for many other types of scientific problems, which included weather prediction, atomic energy calculations, cosmic ray studies, thermal ignition, random-number studies, and wind-tunnel design.

J. Presper Eckert first showed an interest in electronics when he began building radios at the age of five. After earning a degree in electrical engineering from the University of Pennsylvania in 1941, he was offered a graduate fellowship at the Moore School. After ENIAC was completed Eckert and his colleague left their teaching positions there to form the Eckert-Mauchly Computer Corporation.

John W. Mauchly entered Johns Hopkins University as an engineering student, but soon decided that he was more interested in physics. Between 1932, when he was awarded his Ph.D., and his first meeting with Eckert in 1941, Mauchly held the single post in physics at Ursinus College near Philadelphia.

"Faster than a Speeding Bullet"

By 1944 requests for wholly new ballistic tables (demanding the calculation of hundreds of trajectories for each table) were pouring into Aberdeen at a rate of six a day. A skilled person with a desk calculator could compute a single sixty-second trajectory in about twenty hours. The large differential analyzer produced the same result in fifteen minutes. The ENIAC, when completed, would require just half the time of the projectile's flight—thirty seconds— to do the calculations.

John W. Mauchly and J. Presper Eckert Jr. of the Moore School, working with Captain Herman H. Goldstine of the U.S. Army, began planning the ENIAC in 1943, but the machine was not completed until after the war had ended. It was, however, widely used for scientific calculation until the early 1950s.

THE ENIAC
ELECTRONIC NUMERICAL INTEGRATOR AND COMPUTOR

DEVELOPED, DESIGNED AND CONSTRUCTED
· BY THE
MOORE SCHOOL of ELECTRICAL ENGINEERING
OF THE
UNIVERSITY OF PENNSYLVANIA
1944

Captain Herman Goldstine (left) and J. Presper Eckert Jr. with a decade plug-in unit. The ENIAC had a total of two hundred such units, giving it a memory capacity of twenty numbers of ten digits each.

Herman H. Goldstine attributes his early involvement with the development of the computer to "being in the right place at the right time." During a chance meeting at the Aberdeen, Maryland, railroad station, John von Neumann became intrigued with Goldstine's description of the ENIAC project. Later, the two collaborated on the first development of the internally stored program computer.

Ballistics

With the outbreak of World War II in 1939, the U.S. Army set out to improve the differential analyzers used by Aberdeen to calculate ballistic tables. Several features were added which increased their speed and accuracy by a factor of ten, but then a practical limit had been reached for the differential analyzer.

The problem was complicated further by an important characteristic of the ballistic problem—certain factors, such as atmosphere resistance, are defined by numbers, but not by mathematical formulas.

For those calculations, two hundred women were employed, and it was often two to three months before a table could be completed. Mauchly and Eckert saw that a digital calculator could do the job of both differential analyzer and human calculators. On April 2, 1943, they submitted a memo describing an "ELECTRONIC DIFF* ANALYZER" ("DIFF*" stood for difference) that would produce results like those "in the computation of a trajectory by hand." A complete table would be done entirely by machine in only two days.

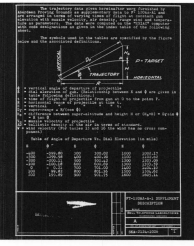

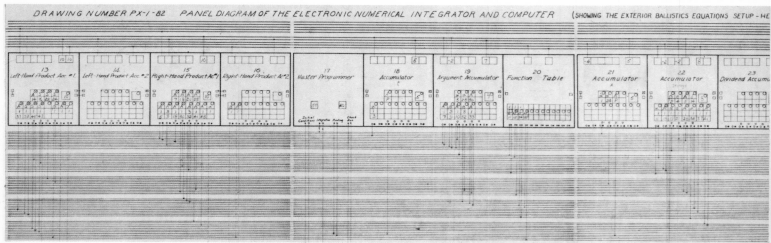

A differential analyzer acquired in 1935 by the Ballistic Research Laboratories at Aberdeen. It was built for the purpose of integrating exterior ballistic equations to produce trajectories for firing and bombing tables.

Programming the ENIAC was accomplished by manually setting switches and reconnecting cables. The chart represents the wiring to set up an exterior ballistic equation.

Some bombing tables at Aberdeen Proving Ground were prepared using IBM relay calculators.

The First Programmers

The building of large, programmable calculators brought with it a new job—programmer. Many of the mathematicians who took on that job were women—partly due to the wartime pressures on manpower, but also to the fact that the newly created field had few precedents for either sex.

Until the early 1950s, many thought that only mathematicians could work with computers, but a clue to what lay ahead was contained in Alan Turing's 1936 paper. By showing that simple computing machines could be instructed to simulate any more complicated machines, he provided a basis for the development of automatic programming.

ADA AUGUSTA
The Countess of Lovelace

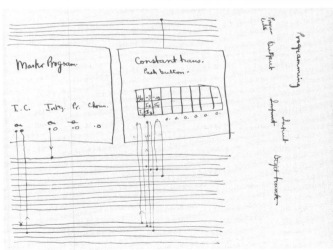

Lord Byron's daughter, Ada Augusta, the Countess of Lovelace, was a skilled mathematician and close friend of Charles Babbage. In a detailed analysis of his proposed "Analytical Engine," she developed the essential ideas of programming.

Mathematician Adele Goldstine, the first programmer on the ENIAC, and her husband, Captain Herman H. Goldstine, who initiated the building of ENIAC for Aberdeen Proving Grounds.

Adele Goldstine's rough notes for the programming of a problem in exterior ballistics. The programming of ENIAC involved setting up the connections by hand—a process that had to be carefully charted in advance.

In World War II, mathematician Grace Hopper was assigned to the Naval Ordnance Computation Project at Harvard. There, as programmer of the Mark I, she developed the original operating programs. Later, working with the UNIVAC, she became a pioneer in the field of computer languages.

Grace Hopper joined the faculty of Vassar, her alma mater, after obtaining a Ph.D. from Yale in 1934. An officer in the Naval Reserve, she became a Harvard Research Fellow at the end of World War II. Later, as senior mathematician for the Eckert-Mauchly Computer Corporation, she wrote the first practical "compiler" program and played an important role in the development of COBOL (COmmon Business Oriented Language).

The von Neumann Concept

In 1945 the three lines of development that had grown from the early calculating machines, statistical machines, and logical automata were converging. New techniques for calculation had been combined with automatic control in gun directors, and with scientific data handling in punch card accounting machines.

One more insight was needed to bring all the ideas and technologies together into the modern computer: the von Neumann concept of the stored program.

136

John von Neumann and the I.A.S. computer. The Institute for Advanced Study had a policy of doing no experimental work. In the fall of 1945 an exception was made so that von Neumann and his group could construct a fully automatic, digital, all-purpose computing machine.

"It is to be expected," von Neumann wrote, "that the future evolution of high-speed computing will be decisively influenced by the experiences gained."

Hungarian-born John von Neumann studied chemistry and mathematics at the universities of Berlin and Budapest and at the Technische Hochschule in Zurich. In 1927 he was appoined privatdozent of the University of Berlin at the unusually young age of twenty-four. He came to the United States in 1930 as a lecturer in mathematical physics at Princeton University, and in 1933 became a permanent member of the Institute for Advanced Study. Although concerned mainly with the theoretical aspects of physics and mathematics, von Neumann's participation in the Manhattan Project gave him an interest in problems of application and convinced him of the great potential of automata. A strikingly learned and hard-working man, he could "talk faster in any of seven languages than most people can in their own."

The Stored Program

In 1944 the Army asked the Moore School to build a more powerful calculator than the ENIAC, which was then still under construction. A year later John von Neumann responded with a complete "logical design" for a machine to be called the EDVAC (Electronic Discrete Variable Automatic Computer).

In von Neumann's paper was imbedded the remarkable idea of a "stored program," now universal to computers. He suggested that the instructions for the computer—always before entered on punched paper tape, or by plugboards— could be stored in the computer's electronic memory as numbers, and treated in exactly the same manner as numerical data.

For the first time, then, logical choices of program sequences could be made *inside* the machine, and the instructions could be modified by the computer as it went along.

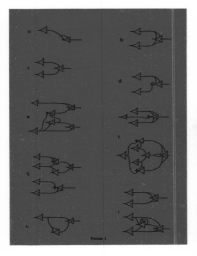

In 1943 Warren S. McCulloch and Walter Pitts developed a notation to illustrate their theory that the "all-or-none" behavior of neural networks can be described in terms of propositional logic. Because of the "on-or-off" character of computer logic, it was a natural notation for von Neumann to elaborate in 1945 when he described EDVAC.

Von Neumann's EDVAC paper of 1945

After the war, Herman Goldstine and Arthur Burks joined von Neumann at the Institute for Advanced Study in Princeton, New Jersey. There they elaborated on von Neumann's 1945 paper to produce a "detailed outline" for the I.A.S. computer and the many others patterned after it.

Arthur Burks, who taught philosophy at Swarthmore College in Pennsylvania and engineering at the Moore School, had earlier collaborated on the ENIAC project.

The principals involved in the building of the I.A.S. computer. From the left, James Pomerene, Julian Bigelow, John von Neumann, Herman Goldstine. Pomerene succeeded Bigelow as chief engineer.

For the memory of their proposed I.A.S. computer, von Neumann, Goldstine, and Burks specified the Selectron tube: "As now planned this tube will have a capacity of $2^{12} = 4096$ binary digits. To achieve a total electronic storage of about 4000 words we propose to use 40 Selectrons, thereby achieving a memory of 2^{12} words of 40 binary digits each."

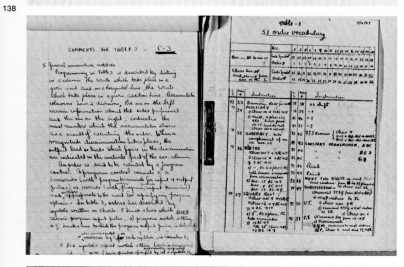

138

(1) A $k+1$-complex : $\overline{X}^{(p)} = (x^0; x^1, \dots, x^p)$ consists of the main number: x^0 and the satellites: $x^1, \dots, x^p$. Throughout what follows $p = 1, 2, \dots$ will be fixed. A complex $\overline{X}^{(p)}$ precedes a complex $\overline{Y}^{(p)}$: $\overline{X}^{(p)} \leq \overline{Y}^{(p)}$, if their main numbers are in this order: $x^0 \leq y^0$.

An n-sequence of complexes: $\{\overline{X}_0^{(p)}, \dots, \overline{X}_{n-1}^{(p)}\}$. If $0', \dots, (n-1)'$ is a permutation of $0, \dots, (n-1)$, then the sequence $\{\overline{X}_{0'}^{(p)}, \dots, \overline{X}_{(n-1)'}^{(p)}\}$ is a permutation of the sequence $\{\overline{X}_0^{(p)}, \dots, \overline{X}_{n-1}^{(p)}\}$. A sequence $\{\overline{X}_0^{(p)}, \dots, \overline{X}_{n-1}^{(p)}\}$ is monotone is its elements appear in their order of precedence: $\overline{X}_0^{(p)} \leq \overline{X}_1^{(p)} \leq \dots \leq \overline{X}_{n-1}^{(p)}$, i.e. $x_0^0 \leq x_1^0 \leq \dots \leq x_{n-1}^0$. Every sequence $\{\overline{X}_0^{(p)}, \dots, \overline{X}_{n-1}^{(p)}\}$ possesses a monotone permutation: $\{\overline{X}_{0'}^{(p)}, \dots, \overline{X}_{(n-1)'}^{(p)}\}$ (at least one). Obtaining this monotone permutation is the operation of sorting the original sequence.

Given two (separately) monotone sequences $\{\overline{X}_0^{(p)}, \dots, \overline{X}_{m-1}^{(p)}\}$ and $\{\overline{Y}_0^{(p)}, \dots, \overline{Y}_{m-1}^{(p)}\}$, sorting the composite sequence $\{\overline{X}_0^{(p)}, \dots, \overline{X}_{m-1}^{(p)}, \overline{Y}_0^{(p)}, \dots, \overline{Y}_{m-1}^{(p)}\}$ is the operation of meshing.

(2) We wish to formulate code instructions for sorting and for meshing, and to see how much control-capacity they tie up and how much time they require. It is convenient to consider meshing first and sorting afterwards.

This 1947 "flow diagram" does not show a real program, but is a sample of all the general features of coding for a computer. The notation, developed by Goldstine and von Neumann, forms the basis for today's flow-diagramming techniques.

Adele Goldstine's "central control" program for ENIAC. Developed before the idea of a "stored program" appeared, ENIAC had to be tediously rewired for each problem. After it had been running for about a year and a half, von Neumann worked with Adele Goldstine to develop a wiring scheme for the ENIAC which allowed it to interpret other programs and act upon them in the same manner as would a "stored program" computer. This modification was completed early in 1948.

The first page of the first program written for a modern computer. Von Neumann had no doubts that the new machine would be effective for scientific computing, so he chose to test its versatility by coding an operation central to business applications —sorting. He illustrated that program, step by step, in an appendix to his June 30, 1945, report on the design of EDVAC.

The Weather Group

Richardson's dream that 64,000 mathematicians could race the weather around the globe, predicting it in advance, was realized when a group headed by John von Neumann used the ENIAC to calculate a twenty-four-hour forecast in twenty-four hours.

In August 1946, von Neumann called a conference to describe a new computer. Saying that he intended to use the machine to forecast the weather numerically, he formed a group at Princeton to tackle that problem.

Jule G. Charney, who joined them in 1948, devised some simplified models of the atmosphere, and by 1950 the group was ready to try a computer run. Since von Neumann's machine was not completed, permission to use ENIAC at Aberdeen Proving Ground was obtained, and the first computer forecast was made.

Jubilant meteorologists in front of the ENIAC at the completion of the first computer weather forecast. Standing from left: Ragnar Fjörtoft, Jule G. Charney, John C. Freeman, and Joseph Smagorinsky. In front are the two programmers from the ENIAC staff that assisted them.

Charney sent the results of the ENIAC run to Richardson. In reply Richardson described a "tiny psychological experiment" in which he asked his wife to decide whether the observed weather at start or the computed weather twenty-four hours later most resembled the observed weather twenty-four hours later. His wife gave a slight edge to the computed forecast.

ENIAC weather forecasts published in *Tellus,* November 1950

The "Analytical Engine"

Charles Babbage's "Analytical Engine," proposed in 1833, had been the standard against which the automatic nature of calculating machines was measured.

Conceived as a giant steam-powered assemblage of brass gears, cams, and wheels, Babbage's machine was never built. In fact, its total concept was not realized until 1946, with the Bell Laboratories Model V. The designer, George Stibitz, was not aware of Babbage's work.

George Stibitz

After receiving his Ph. D. from Cornell University in 1930, Stibitz was hired by Bell Telephone Laboratories as a research mathematician. He left in 1941 to spend the war years as a technical aide to the National Defense Research Council. His contributions to the computer field include the development of binary and floating-point arithmetic.

Model V relay calculator installed at Aberdeen Proving Ground. The first Model V was delivered to Langley Field, Virginia, in December 1946. Built from parts found in ordinary telephone systems, they required only simple maintenance. Consequently, their reliability was outstanding—the Model V worked an average of twenty two hours a day.

Stibitz Model V

In the design of his "Analytical Engine" Babbage listed the four elements a machine had to have to perform the functions of a human computer: an arithmetic unit; a "memory"; automatic "choice" of computing sequence; input and output.

George Stibitz designed five models of relay calculators. Models I to IV did arithmetic and had memories and inputs and outputs. But the Model V incorporated the critical feature of "choice"; based on the result of its own calculation, it automatically chose one of several computing sequences to do next.

Information was entered into Bell Laboratories Model I calculator via three teletype terminals. "There was a certain amount of sneaky behavior on the part of the operators trying to get to a teletype and turn it on before the other teletypes could be occupied," Stibitz later recalled, "because what we had then for time-sharing did not allow two people to work at the same time."

Bell Laboratories Model III relay calculator, delivered in 1944 to the Army's Anti-aircraft Artillery Board

Teletype tape and reader. Beginning with the Model II, the Bell Laboratories relay calculators used teletype tape punched in a five-hole code for input and output. The punched tapes contained data, the machine's program, and special tables.

Operations Research

In World War II, the number and complexity of new technologies compelled first the British, then the Americans, to call in scientists who could advise on operational matters.

Working in such areas as radar air defense, convoy deployment, and antisubmarine strategies, these scientists developed a body of techniques and attitudes that became known as "operations research."

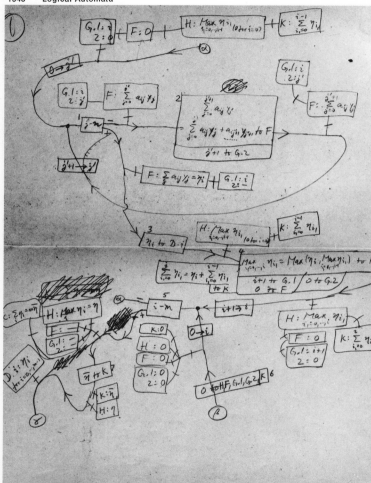

John von Neumann

Oskar Morgenstern

A page from von Neumann's working notes. During the planning of the I.A.S. machine, he considered many kinds of problems for the computer. This one, in game theory, was formulated by E. W. Paxson in 1946, and involved a duel between a destroyer and a submarine. Each could gather data about the other with sonar; if the submarine could get into the destroyer's wake, it would be undetectable.

In the progress of a day's work with Paxson, von Neumann posed an abstract solution to the problem, indicated the form of the programming flowcharts, and even wrote some typical computer code.

In their book *Theory of Games and Economic Behavior* John von Neumann and Oskar Morgenstern analyzed certain simple games, such as poker and coin matching, to show that there is a mathematically determinable "best possible" method of play, regardless of what the opponents do. They showed that the theory was applicable "on the one hand to games in the proper sense, on the other to economic and sociological problems."

Linear Programming

At the end of the war, the U.S. Air Force initiated Project SCOOP (Scientific Computation of Optimum Programs) to mechanize and speed up the planning and deployment process.

In 1947 George Dantzig was working on this project to build an abstract model of the planning process. In generalizing the linear programming scheme Wassily Leontief had developed for solving input-output economic models, Dantzig devised the "simplex method," which became the basis for the most widely used techniques of computing best strategies.

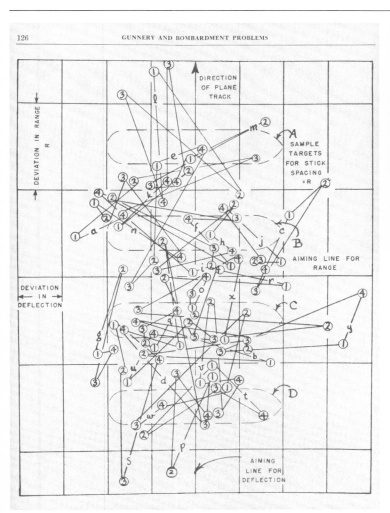

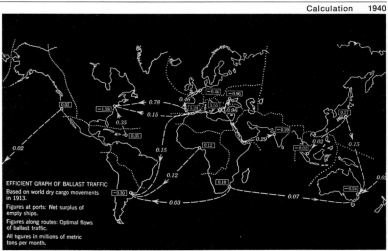

EFFICIENT GRAPH OF BALLAST TRAFFIC
Based on world dry cargo movements in 1913.
Figures at ports: Net surplus of empty ships.
Figures along routes: Optimal flows of ballast traffic.
All figures in millions of metric tons per month.

143

During World War II, a shortage of cargo ships created critical supply bottlenecks for the Allies. Tjalling Koopmans approached the situation as an optimum scheduling problem; this allowed him to provide "best" solutions to ship routing by linear programming methods. After the war, Koopmans spearheaded research into the potential of applying these same methods to problems in economics.

Information Theory

Stimulated by Norbert Wiener and formalized in 1948 by Claude E. Shannon, information theory treats communication as a problem in statistics. The theory permits precise measurement of the amount of information delivered and the efficiency of machines that handle it.

Like computers, the theory deals with "information" rather than meaning. But physicist Warren Weaver commented that the theory "has so penetratingly cleared the air that one is now, perhaps for the first time, ready for a real theory of meaning."

"There is no reverse on a motorcycle a friend of mine found this out rather dramatically the other day." Shannon asked people to guess, one by one, the letters in the above sentence. With such guessing games, he determined just how much "information"

1940   Statistics

Claude E. Shannon

144

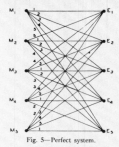

COMMUNICATION THEORY OF SECRECY SYSTEMS   681

into a given cryptogram E is equal to that of all keys transforming M_j into the same E, for all M_i, M_j and E.

Now there must be as many E's as there are M's since, for a fixed i, T_i gives a one-to-one correspondence between all the M's and some of the E's. For perfect secrecy $P_M(E) = P(E) \neq 0$ for any of these E's and any M. Hence there is at least one key transforming any M into any of these E's. But all the keys from a fixed M to different E's must be different, and therefore *the number of different keys is at least as great as the number of M's.* It is possible to obtain perfect secrecy with only this number of keys, as

Fig. 5—Perfect system.

one shows by the following example: Let the M_i be numbered 1 to n and the E_i the same, and using n keys let

$$T_i M_j = E_s$$

where $s = i + j$ (Mod n). In this case we see that $P_R(M) = \frac{1}{n} = P(E)$ and we have perfect secrecy. An example is shown in Fig. 5 with $s = i + j - 1$ (Mod 5).

Perfect systems in which the number of cryptograms, the number of messages, and the number of keys are all equal are characterized by the properties that (1) each M is connected to each E by exactly one line, (2) all keys are equally likely. Thus the matrix representation of the system is a "Latin square."

In MTC it was shown that information may be conveniently measured by means of entropy. If we have a set of possibilities with probabilities p_1, p_2, $\cdots$, p_n, the entropy H is given by:

$$H = -\sum p_i \log p_i.$$

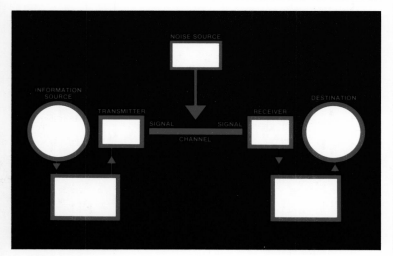

A schematic diagram of a general communication system. The information source produces a message which is encoded, then sent by the transmitter through a channel to the receiver. There it is decoded and delivered to its destination. The amount of error in the message depends on the amount of interference ("noise") during transmission.

Information theory, in one of its most surprising proofs, shows that, despite any

amount of noise, a given communication channel can be made to carry information up to its capacity with negligibly small error—perfect transmission from an imperfect system.

Information theory is so general that it includes the whole of cryptography. During the war, Shannon worked on codebreaking techniques at Bell Laboratories. Starting about 1941, he worked on both information theory and secrecy systems. "I wouldn't say one came before the other," he said. "They were so close together you couldn't separate them."

Both the Americans and the British built sophisticated special-purpose electronic

calculating equipment for codebreaking during World War II. When dismantled at the war's end, one of the British machines became the frame for the "baby" machine, prototype of the Manchester computer, MADM.

Cryptography

is conveyed by each letter in English.

In Shannon's information theory, the more difficult it is to guess what the next letter or symbol in a message will be, the more "information" the message contains. Conversely, the more easily the receiver of the message can guess the next symbol, the less "information." This idea gives the engineer a way to exploit the statistical character of communication. The part of the message he can predict, he need not transmit.

Secret codes and ciphers had been, for hundreds of years, an isolated specialty. But, in 1922, William Friedman published a brilliant paper connecting cryptography with the rich field of mathematical statistics: "The Index of Coincidence and its Applications in Cryptography." His insights led to a wealth of new codebreaking techniques, and, in the 1930s, to punch card installations at Pearl Harbor, Corregidor, and Washington. The success of these methods proved a vital factor in World War II.

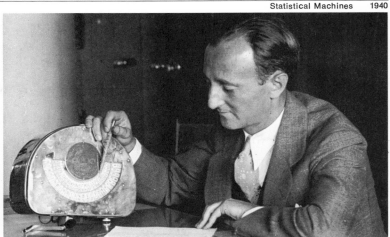

William Friedman demonstrating a cipher machine he invented

Who Says a Watched Pot Never Boils?

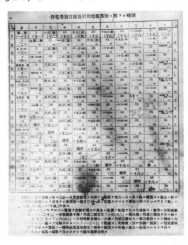

Gray code disk. Codes are used in communication primarily to reduce error or to improve the efficiency of transmission. This disk carries a code designed to reduce errors in sensing a message at the receiving end. Called a "cyclic permutation code," the Gray code is an arrangement of information represented as 1's and 0's in such a way that each group varies from the group beside it by no more than one digit.

Because of his fame as a codebreaker, Friedman was called upon to solve the relatively simple code used in the Teapot Dome case in 1924. He testified that this message to Edward McLean —"ZEH HOCUSING IMAGERY COMMENSAL ABAD OPAQUE HOSIER LECTIONARY STOP CLOT PRATTLER LAMB JAQUAR ROVED TIMEPIECE NUDITY"— actually read "Zerely thinks trend of investigation favorable to you. Not impressed with Walsh as cross-examiner."

Japanese code book. Friedman's statistical methods encouraged the introduction of tabulating machines into cryptography during the 1930s, and the building of special-purpose all-electronic computers during World War II.

In a dramatic application of these methods, his team painstakingly broke the most secret and complex Japanese cipher, the PURPLE cipher, by constructing a duplicate of the rotor mechanism that produced it.

Cybernetics

The automatic machines that developed out of World War II technology used self-regulating mechanisms to replace human control. They performed such difficult tasks that they appeared to exhibit qualities peculiar to the human brain. In 1948 Norbert Wiener captured the essence of these considerations in his book, *Cybernetics: or Control and Communication in The Animal and The Machine.*

At the same time, John von Neumann was developing a theory of automata that concentrated on the similarities between the human brain and electronic computers.

William Grey Walter became head of the Physiology department at Burden Neurological Institute, Bristol, England, in 1939. A student of the electrical activity and diseases of the brain, he discovered delta and theta brain waves.

In 1948 W. Grey Walter built an electromechanical tortoise to study simple reflex actions, especially the idea that complex behavior depends more on the richness of interconnections than on the number of original elements. "These machines are perhaps the simplest that can be said to resemble animals," Walter wrote. "Crude though they are, they give an eerie impression of purposefulness, independence and spontaneity."

Grey Walter with his tortoise in its traveling case (which also serves as its recharging station).

The path taken by the tortoise to reach its hutch. If its batteries are run down, the tortoise will go into the hutch to be recharged. Otherwise the tortoise is attracted to the strong light at first, but then repelled when it gets too close.

Theory of Automata

The emergence of electronic computers spurred speculation about the future of "thinking" machines and automation. Norbert Wiener gave a name to the field of computer control with his book *Cybernetics,* in which he explored the potential uses of such automata.

Von Neumann's theory of automata grew out of the development of computers. Its purpose was to draw conclusions about complex natural organisms based on experience gained with artificial automata. But he hoped, in turn, that the theory would contribute to the development of very complicated computers.

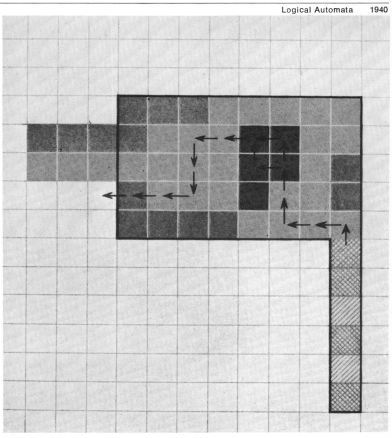

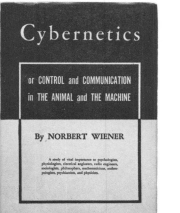

John von Neumann

Norbert Wiener was born in Columbia, Missouri, in 1894, the son of a professor of modern languages. A child prodigy, he graduated from high school at twelve and was awarded a Ph.D. in mathematical logic by Harvard at nineteen. In 1919 he began teaching at the Massachusetts Institute of Technology.

Diagram of von Neumann's "self-reproducing automata" from *Scientific American,* April 1955.

Von Neumann built on the idea of the Universal Turing Machine, and sought to extend Turing's theoretical limit on the behavior of any arbitrary automaton to include self-reproduction.

He described a machine, a combination of brain and brawn, that could instruct itself to build machines to fit a certain description. If that description is identical to the machine itself, one could say it is self-reproducing.

In order to visualize how his machine would be built, von Neumann bought the biggest Tinker Toy set that money could buy and began producing individual parts—until he realized he could represent the machine in two dimensions. Shortly afterward, his "Theory of Games" colleague, Oskar Morgenstern, received a giant present for his son.

Automation

While Leonardo Torres had advocated the automatic control of industrial machines in 1913, the term "automation" was actually coined in 1947.

1940   Logical Automata

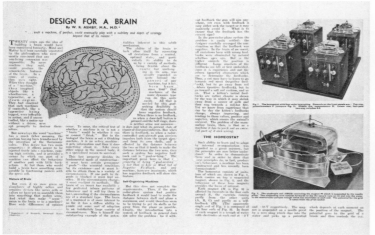

W. Ross Ashby

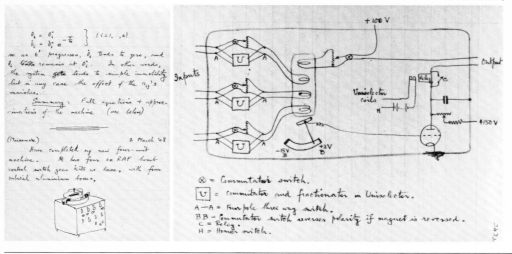

In a series of notebooks, W. Ross Ashby carefully recorded his ideas about the mechanistic nature of the human brain and his attempts to imitate some of its fundamental activities. In this volume (number 11 of 24) he diagrammed the machine he was building, and eleven days later made this entry: "Triumph! The machine of p. 2432, the 'automatic homeostat,' was completed today after some wrong wirings, a burnout due to shorting, delay for fitting fuses everywhere and finally an actual working this evening. After all the trouble it works." These notebooks formed the basis for Ashby's *Design for a Brain* (1952).

Ashby created his "homeostat" to study the mechanisms by which animals preserve their internal stability. Norbert Wiener wrote that "such a machine can learn to a limited degree: that is, it can adjust itself by its mode of behavior to a stable balance with its environment."

At Ford Motor Company, Delmar S. Harder formed the Automation Department to find ways to use self-regulating machines to run production. In 1949 Ford began work on the first factories built specifically for automation. When completed, Ford's Cleveland Engine Plant and Foundry reduced the time required to make a cylinder block from 9 hours to 14.6 minutes.

Computers and Chess

The game of chess has often seemed an ideal test of a computing machine's abilities. Charles Babbage thought his Analytical Engine of 1833, if completed, would have been able to play chess. Leonardo Torres built an end game chess machine in 1914 to prove his theory that machines could do many things "popularly classed as thought."

In 1947-48 Alan Turing and David Champernowne wrote complete specifications for a one-move analyzing chess machine they named the "Turochamp." Donald Michie and Shaun Wylie designed a rival "paper machine," the "Machiavelli." A playoff was never completed—the hand calculations required for each team to simulate their machines were too laborious. When the MADM computer was finished, Turing programmed it to be the first machine capable of playing a complete game of chess.

"If one could devise a successful chess machine, one would seem to have penetrated to the core of human intellectual endeavor."

— Allen Newell, J. C. Shaw H. A. Simon, 1958

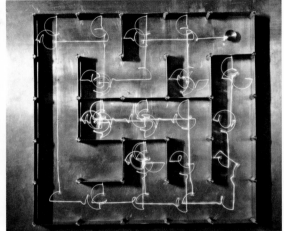

To try out his ideas about programming a computer to play chess, Claude Shannon built a machine to play a variety of end games—actually, an advanced version of Torres' chess player.

Shannon's machine would compute the advantages of various possible moves. It would then signal its move by flashing the light in the appropriate squares of this playing board. It was named "Caissac," after Caïssa, Muse of Chess.

In 1949 Shannon analyzed the problems of programming a computer to play chess in a paper which established a framework for most of the subsequent work in this field. He felt that chess was useful to study machine strategies because it is a problem with "a real form without the complexities of the real world."

Shannon built his maze-solving mouse to study a different labyrinthian problem—telephone switching systems. Like the mouse, a telephone call must make its way to its destination by the shortest possible path.

He described his machine as being "capable of solving a maze by trial-and-error means, of remembering the solution, and also of forgetting it in case the situation changes and the solution is no longer applicable."

Display panel from Shannon's "mouse." The red light indicated that the mouse was present in the square; the white lights displayed the path followed as it made its way through the maze.

Any of the barriers in the maze could be changed. The mouse would then retrace without error any part of the maze that remained the same and solve the altered parts by trial and error.

Simulation in Real Time

In June 1946, M.I.T.'s Servomechanisms Laboratory redefined a U.S. Navy project from one which called for the design of an aircraft simulator to one which set out to build an electronic computer.

Named "Project Whirlwind," it was "for use in control and simulation work such as air traffic control, industrial process control, and aircraft simulation."

As an engineering model, Whirlwind was perhaps the most influential of the early computers in terms of today's commercial machines. Both the invention of "coincident-current" magnetic core memory and the parallel synchronous method for handling information inside the machine were first developed by Whirlwind's designers.

"Whirlwind" and Control

Jay W. Forrester and his group at M.I.T. were familiar with the ENIAC and von Neumann's plans for the EDVAC. Nevertheless, Whirlwind I, reflecting its designers' background in automatic control engineering, was developed along different lines of technology than the other computers of the late 1940s.

Whirlwind's commitment to "real time" control led to its use as the prototype of the SAGE air defense system. The system was successfully simulated at Cape Cod in 1951 by linking radars to Whirlwind. The computer automatically identified unfriendly aircraft, predicted their courses, and directed interceptor fighters.

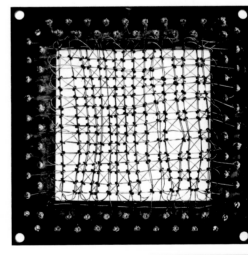

Jay Forrester (standing), Patrick Youtz, and Stephen H. Dodd examine some of the electrostatic storage tubes that comprised Whirlwind's initial memory.

One of the original four magnetic cores developed by Jay Forrester in 1949. Wire-wrapped, it successfully operated as experimental memory in a two-by-two array.

Jay Wright Forrester was born on a Nebraska cattle ranch in 1918. As a graduate student at M.I.T., he worked with Gordon S. Brown, founder of M.I.T.'s Servomechanisms Laboratory. From 1945 to 1952 Forrester was the director of the Digital Computer Laboratory, which built the Whirlwind.

Airmen operating the consoles of Whirlwind in a test for the SAGE (Semi-Automatic Ground Environment) system.

The computer identified hostile aircraft by comparing the information picked up on radar with the flight plans of friendly planes stored in its memory. The SAGE system became operational in 1958 after 1800 man-years of computer programming.

One of the sixteen-by-sixteen ceramic core arrays built to prove the idea of magnetic core storage. The first complete core memory actually used in Whirlwind consisted of sixteen larger "core planes," each capable of storing 1024 binary digits.

S.S.E.C.

At the end of the war, the Mark I relay calculator was running at Harvard, the all-electronic ENIAC was nearing completion at the Moore School, and John von Neumann was developing the stored program concept.

IBM then made a decision to build a machine that combined existing mechanical and electronic technologies to provide a very large number of inputs and a form of stored program. The first machine that could control its calculating

sequence by modifying its own instructions, it was named the Selective Sequence Electronic Calculator.

Wallace J. Eckert

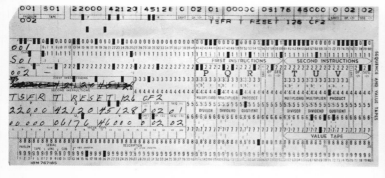

T. J. Watson Sr. designated W. J. Eckert, director of IBM's Watson Scientific Computing Laboratory, as the "internal customer" for the building of the S.S.E.C. Eckert used it for, among other things, calculations of the moon's orbit which would be included in the Apollo program twenty years later. The instruction card is from the program for those calculations.

The operation indicator and control desk of S.S.E.C. The operator could follow the progress of the calculations and change the procedure if necessary.

The S.S.E.C. printed out calculations in progress, as well as the final results.

THE SATURDAY EVENING POST   December 16, 1950

INTERNATIONAL BUSINESS MACHINES, famous name in office equipment, builds some of the world's most complex and efficient machines. Shell Industrial Lubricants are used in many operations.

Oracle on 57th Street

"This machine will assist the scientist in institutions of learning, in government, and in industry to explore the consequences of man's thought to the outermost reaches of time, space, and physical conditions."

Thos. J. Watson
(Inscription on the S.S.E.C.)

From a Shell Oil Company advertisement in *The Saturday Evening Post*, December 1950.

Perhaps the first giant digital calculator available for commercial use, the S.S.E.C., during the four years it was in operation, undertook calculations for a variety of customers. General Electric posed a problem on the design of turbine buckets, and Gulf Research and Development Company and Carter Oil Company both requested solutions to oil field exploration problems.

The S.S.E.C was formally dedicated on January 28, 1948, at IBM's New York showroom, 57th Street and Madison Avenue.

The Computer

In his design for the EDVAC, John von Neumann had given a "blueprint" for the modern computer; almost immediately, a number of machines were started in Britain and the United States.

Most of the new machines were named in the acronymic style of the ENIAC—EDVAC, EDSAC, MADM, UNIVAC, SEAC, SWAC, MANIAC, NORC; perhaps fifteen in all were in progress or completed by 1950.

Although all these computers were modern, electronic, digital, stored-program machines, their technologies differed considerably. This was possible because the 1946 report of Burks, Goldstine, and von Neumann, the "detailed outline" for the Institute for Advanced Study computer, was written in terms of a logical design (an "architecture"). This important idea allowed a variety of technological solutions.

154

THEORY AND TECHNIQUES

FOR DESIGN OF

ELECTRONIC DIGITAL COMPUTERS

Lectures given at the Moore School
8 July 1946 — 31 August 1946

Volume I
Lectures 1-10

UNIVERSITY OF PENNSYLVANIA

Moore School of Electrical Engineering

PHILADELPHIA, PENNSYLVANIA

September 10, 1947

ENIAC had been running for less than a year when, in the summer of 1946, the Moore School was the host for a course on computer design and the von Neumann concept. The students were twenty-nine professionals drawn from twenty British and American organizations, and the lectures served as a catalyst for the development of the modern computer.

ENIAC—Electronic Numerical Integrator And Computer
EDVAC—Electronic Discrete Variable Automatic Computer
EDSAC—Electronic Delay Storage Automatic Calculator
MADM—Manchester Automatic Digital Machine
UNIVAC—UNIVersal Automatic Computer
SEAC—Standards Eastern Automatic Computer
SWAC—Standards Western Automatic Computer
MANIAC—Mathematical Analyzer, Numerical Integrator And Computer
NORC—Naval Ordnance Research Calculator

EDSAC

Appropriately enough, the first of the modern computers completed was built in England, where Babbage had described his Analytical Engine in 1833, and Turing the theory basic to all computers in 1936. Constructed by Maurice V. Wilkes at Cambridge University, EDSAC did its first automatic computation on May 6, 1949.

The EDSAC had oscilloscope tubes for displaying stored information. The center one shows the contents of the mercury delay memory.

Wilkes worked on the EDSAC design while attending the lectures at the Moore School and during his return voyage on the Queen Mary. Eager to have an operating machine as soon as possible, he deliberately made the design simple.

The control and computing units of the EDSAC used short acoustic delay tubes to store numbers for short periods of time (1 minor cycle, and ½ minor cycle delays).

According to Wilkes, during a 1947 discussion on what fluid to use in such tubes, Alan Turing had advocated the use of gin, which he said contained alcohol and water in just the right proportions to give a zero temperature coefficient of propagation velocity at room temperature.

MADM

An experimental computer built by
F. C. Williams and T. Kilburn
began successful operation at
Manchester University, England,
in June 1949.

SATURDAY, JUNE 25, 1949.

A CALCULATING
MACHINE WITH A
"MEMORY": THE
CONTROL PANEL, AND
A STORAGE TUBE IN USE.

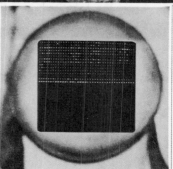

The use of a cathode ray tube to store
information had been considered and
rejected by others, but in 1948, Williams
proved at Manchester that such tubes
could be used as a kind of fast
"delay line."

I.A.S.

The I.A.S. Computer, built at the Institute for Advanced Study at Princeton, was started in 1946 and completed in 1951. A number of computers were modeled after it, including ORACLE, JOHNNIAC, MANIAC, and ILLIAC I.

At the same time that von Neumann was building the I.A.S. computer, the Atomic Energy Commission began a machine based on von Neumann's plans: the MANIAC.

157

James Pomerene with one of the forty Williams tubes used for "random access" memory in the I.A.S. computer. Each tube could store 1024 "bits" of information.

UNIVAC

In 1947 large-scale calculating machines were for the most part thought of as expensive scientific specialty items. Yet that year the principals in ENIAC's design formed the Eckert-Mauchly Computer Corporation to build the first computer designed for commercial use—the UNIVAC.

158

Iron-plated metal tape was used to put data and instructions into the UNIVAC. As the operator typed the information, electric pulses magnetized a pattern of dots across the tape.
Lent by the R.E.S.I.S.T.O.R.S.

Although the first three were ordered by the government, UNIVAC had been developed by Eckert and Mauchly expressly for the business world. Studies were made of typical applications in insurance, a field with perhaps the largest requirements for data processing. In 1949, two years before any UNIVACs were completed, Grace Hopper programmed a typical mortality study and the calculation of extended insurance shown here.

In the late 1940s, about 10,000 tons of punch cards per year were being used in the United States. The builders of the UNIVAC adopted iron-plated metal tape as a way to store large quantities of information, by using electric pulses to magnetize a pattern of dots across the tape. The use of tape as a compact means of storage in the 1950 census led quickly to its use by commercial firms.

The first UNIVAC was delivered to the Bureau of the Census on June 14, 1951. J. Presper Eckert is seated at the console; standing, from left: James H. Rand, president of Remington Rand; Roy V. Peel, director of the U.S. Census; Lieutenant General Leslie R. Groves, vice president of Remington Rand.

SEAC

The Air Force at first supported the development of the UNIVAC for use in speeding up their planning and deployment (Project SCOOP); but when in 1948 it became apparent that the machine would not be finished in time, Air Force officials asked the National Bureau of Standards to undertake a crash program to build a computer. It became SEAC, which began operation in April 1950.

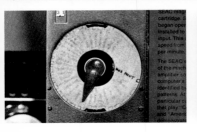

Shortly after SEAC began operation, magnetic wire was installed to replace punched tape as input. This increased the reading speed from 30 to 10,000 words per minute.

The SEAC's operators attached one of the machine's registers to an amplifier and speaker so that the malfunctions in the computer's routines could be identified by unfamiliar sound patterns. At the end of this particular program cartridge there are programs that play "Camptown Races," "Dixie," and "America"—a diversion demonstrating the programmers' skills.

SEAC power control panel. The "running time" clock indicates 70,254 hours—the total time the SEAC ran between April 1950 and its final day of operation, October 20, 1964.

Acoustic delay line. The memory in SEAC consisted of sixty four mercury-filled glass tubes with a quartz crystal at each end (one as a transmitter and the other as a receiver). Each acoustic delay line had a capacity of eight words—a word being a sequence of information bits in the form of sound waves traveling through mercury.

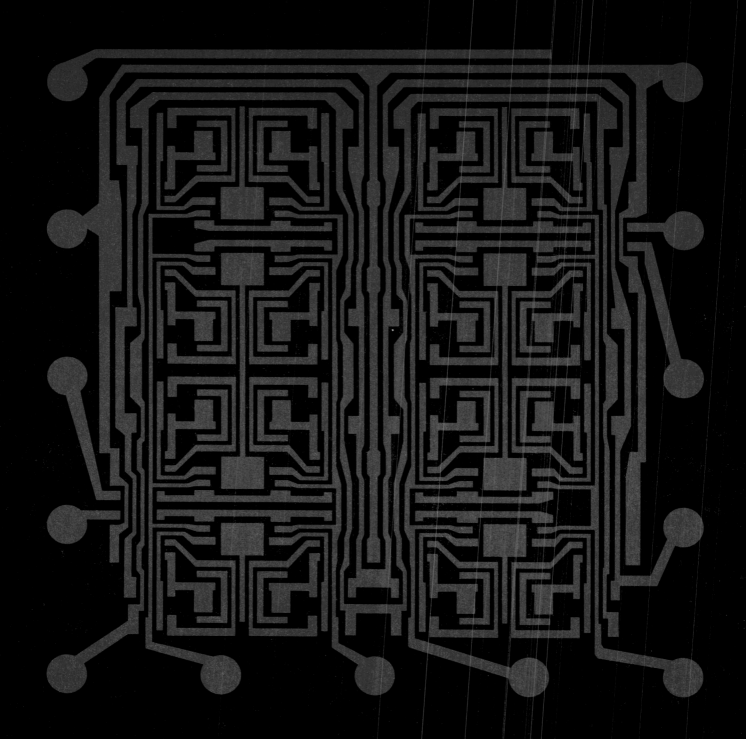

By 1950 the computer had appeared in essentially its present form. But even those closest to it were unprepared for what would follow. The computer's spectacular growth—in numbers, in capability, in application—came as one of the great surprises of modern times.

What may not be surprising, but can be heartening, is that the computer appears to be the result of many people trying to solve many problems in many fields—as a natural consequence of getting on with the business of life in general.

1950-1955

1955-1960

The idea of the computer as a commercial product was beginning to take shape.

In 1951 Remington Rand installed the first production line computer, UNIVAC I, in the Bureau of the Census.

By the end of 1953, thirteen companies were manufacturing computers; IBM and Remington Rand led the field with a combined total of nine installations.

By 1954 it was estimated that perhaps fifty companies in the country could eventually use "electronic brains." The same year the first computer for commercial applications, a UNIVAC I, was delivered to General Electric's new appliance plant in Louisville, Kentucky.

The predominant technologies were cathode-ray tubes and magnetic drums for computer memories, and vacuum tubes for logic and arithmetic.

One measure of available computing "power" is the number of additions that all computers installed in the United States could perform in one second. For 1955, this number is estimated at 250,000/sec.*— about the same speed of operation as a single fairly large computer today.

The development of computer "languages"—programs which remove some of the tedium of programming computers—led to a significant growth in the number of users.

The number of installations increased to approximately 6000 by the end of the decade.

In 1958, after six years of development and 1800 man-years of programming effort, the complex SAGE Air Defense System became operational.

The first fully automated computer-based system for process control was installed at Texaco's Port Arthur Texas Refinery in 1959. That same year, the banking industry adopted MICR (Magnetic Ink Character Recognition).

Ferrite cores became the standard for memories, and transistors replaced vacuum tubes. Total computer power increased by a multiple of 20* between 1955 and 1960.

162

*Asterisk indicates estimates provided by the American Federation of Information Processing Societies (AFIPS)

Computers came into general use in scientific and business operations; more exotic applications were anticipated.

In 1960 Bethlehem Steel became the first company to use its computer, a UNIVAC, to handle orders, inventories, and production control on a real time basis. The Bank of America installed automatic check-processing equipment.

In 1963 the *Daily Oklahoman-Oklahoma City Times* became the first newspaper to set all editorial and classified copy by computer. The American Airlines/IBM real time seat reservation system, SABRE, was made fully operational in 1964.

By 1964 the idea of a new generation of machines—integrated circuits and a fully compatible family of computers—was established with the IBM System/360.

Faster circuitry, and a continuing growth in the number of installations between 1955 and 1965, increased the total power of all computers in the United States 800-fold.*

The growth of the computer industry was reflected in the announcements of some two hundred new machines by over eighty major manufacturers.

New ways were devised for getting information into and out of computers. In 1965 the New York Stock Exchange installed a computer which gave voice answers to telephone queries.

Direct employment for programmers, operators, manufacturing and sales and service personnel was around one million* by 1970. Perhaps an additional half million used computers on an occasional or part-time basis.

Miniaturized circuits, with speeds of operation measured in billionths of a second, contained over 1400 elements—transistors, resistors, and diodes—on chips an eighth of an inch square.

Calculating speeds increased dramatically. The reduction in computation costs (by a factor of 10*) encouraged many new uses—and users—of the information machine. New programming systems increased programmer productivity and made the computer easier to use. By the end of the decade, computers were performing tasks in more than 3,000 different categories.

163

This book is based on the exhibition

A Computer Perspective

Designed by the Office of
Charles and Ray Eames for IBM

Story Glen Fleck
Exhibition Robert Staples
Research Barbara Fahs Charles
Production David Olney/Karl Rimer

Consultants
I. Bernard Cohen/Owen Gingerich/
Ralph Caplan

The assistance of IBM employees
in the development of the exhibition
and this book is especially
acknowledged.

The help and
cooperation of the
following individuals,
organizations and
institutions are
gratefully
acknowledged

**For Artifacts, Aid
and Consultation**
Charles W. Adams
W. Ross Ashby
Preston R. Bassett
L. Beuken
Julian H. Bigelow
Geraldine Binet
Elizabeth Bollée Binford
Harold S. Black
John H. Blodgett
David C. Bomberger
Alain Brieux
Gordon S. Brown
Theodore H. Brown
Arthur W. Burks
Robert Burns
Vannevar Bush
H. R. Calvert
Jule G. Charney
Perry Crawford
George Dantzig
Edith Dimmett
Stuart C. Dodd
Paul Donaldson
Wallace J. Eckert
M. H. Fisch
Sir David Follett
Jay W. Forrester
Herman H. Goldstine
I. J. Good
Arthur Griffin
L. C. Hallman
Edwin L. Harder
Grace Hopper
George Howard
Claude A. R. Kagan
Wassily W. Leontief
Clarence A. Lovell
J. P. Madden
Ethel C. Marden
D. T. McAllister
John C. McPherson
Donald H. Menzel
David B. Parkinson
David M. Parkyn
E. W. Paxson
E. S. Pearson
George W. Platzman
Arthur Porter
John Rafferty
Stephen A. Richardson
Walter Rosenblith
Claude E. Shannon
B. F. Skinner
Joseph Smagorinsky
George Stibitz
Leonardo Torres-Quevedo
Elaine Traylen
John W. Tukey
W. Grey Walter
Warren Weaver
Marina Whitman
Margaret E. Wiener
John V. Wilbur
Maurice V. Wilkes
Ben D. Wood
H. Zemanek

For the Loan of Machines and Materials
Academy of Motion Picture Arts and Sciences
Aktiebolaget Original-Odhner, Sweden
American Federation of Information Processing Societies
American Institute of Public Opinion, The Gallup Poll
American Telephone and Telegraph Company
Archives of the History of American Psychology,
    University of Akron
Armed Forces Institute of Pathology
Bell Telephone Laboratories
Bunker-Ramo Corporation
California Institute of Technology
University of California at Los Angeles
Cambridge University Press
The Carnegie Foundation for the Advancement of Teaching
Bureau of the Census, Department of Commerce
University of Chicago Library
Department of Physical Sciences, University of Chicago
Yerkes Observatory, University of Chicago
CBS Television Network
Columbia University
Columbiana Collection, Columbia University
Conservatoire National des Arts et Métiers, Paris
L'Ecole des Arts et Métiers, Paris
Educational Testing Service
Escuela de Caminos, Madrid
Ford Archives, Henry Ford Museum
Film Library, Ford Motor Company
Harrah's Automobile Collection
Graduate School of Business Administration,
    Harvard University
The University of Iowa Libraries
King's College, Cambridge
The Library of Congress
Link Division of the Singer Company
Lowell Observatory
The State Library, Commonwealth of Massachusetts
Massachusetts Institute of Technology
Metropolitan Life Insurance Company
University of Michigan
The MITRE Corporation
Museum of Modern Art Film Stills Archives
National Archives
National Bureau of Standards
National Library of Medicine
H. M. Nautical Almanac Office,
    Royal Greenwich Observatory, England
New York Life Insurance Company
The New York Public Library
New York Times
The New Yorker Magazine, Inc.
NOAA, National Ocean Survey, Department of Commerce
Parke, Davis & Company
Peabody Museum, Harvard University
The General Alumni Society, University of Pennsylvania
The Library and Museum of the
    Performing Arts at Lincoln Center
Philadelphia Evening Bulletin
Pioneer Hi-Bred Corn Company
Princeton University Library
The Prudential Insurance Company of America
R.E.S.I.S.T.O.R.S.
The Science Museum, London
Smithsonian Institution, National Air and Space Museum
Smithsonian Institution,
    National Museum of History and Technology
Social Security Administration,
    Department of Health, Education and Welfare
University of Southern California
Technocracy, Inc.
The Galton Laboratory, University College London
University College London
University Library of Oslo
U.S. Army Aberdeen Research & Development Center,
    Aberdeen Proving Ground
U.S. Naval Academy Museum
U.S. Government Patent Office
Zuse KG. Germany

165

Acknowledgments

1	2	3	4
5	6	7	8
9	10	11	12
13	14	15	16

Text and photograph acknowledgments are keyed numerically to a 16 square matrix of each page. Information is listed in this section by page number and position number.

Illustrations not referenced are from the IBM archives and antique calculator collection.

Page 12   13: C. Babbage, *On the Economy of Machinery and Manufactures* (London, C. Knight, 1832)

Page 13   5: C. Babbage, *Babbage's Calculating Engines* (London, Spon, 1889), foldout.

Page 14   7a: Science Museum, London. 9: U.S. Patent #159, 244 (1875). 12a: Conservatoire National des Arts et Metiérs, Paris, France. 12b: Science Museum, London.

Page 15   5: *Royal Astronomical Society Occasional Notes* 2: plate 5, (Aug. 1847). 6: *Yearbook of Facts* (London, Timbs, 1847). 7: Science Museum, London. 8: "Our Portrait Gallery," *Dublin University Magazine* 89 (Jan.-Jun. 1877), 2nd ser. #40. 9: *L'Illustration* 4:156 (1846). 12: Science Museum, London.

Page 16   5: Charles Darwin, *On the Origin of the Species by Means of Natural Selection* (Chicago, Encyclopedia Britannica, 1952) p. 56. 6a, 12: Culver Pictures, Inc., New York, N.Y. 6b: James R. Newman, *What is Science?* (New York, Simon and Schuster, 1955), p. 436. 7: Helen M. Walker, *Studies in the History of Statistical Method* (Baltimore, Williams and Wilkins, 1929). 8: Adolphe Quetelet, *Physique sociale: Essai sur le Développement des facultés de l'homme* (Brussels, C. Muquardt, 1869), cover; Department of Applied Statistics, University College London, Eng. 9, 10: Karl Pearson, *The Life, Letters and Labours of Francis Galton, IIIa: Correlation, Personal Identification and Eugenics* (Cambridge, Cambridge University Press, 1930). 11: Florence Nightingale, *Notes on Matters Affecting the Health, Efficiency, and Hospital Administration of the British Army Founded Chiefly on the Experiences of the Late War* (London, Harrison and Sons, 1858).

Page 17   5: Parke Davis and Co., Detroit, Mich. 7, 12: Bureau of the Census, U.S. Government, Washington, D.C. 8: D. R. Dewey, *Discussions in Economics and Statistics of Francis A. Walker* (New York, Holt, 1899), frontispiece. 9: Warshaw Collection, Smithsonian Institution, Washington, D.C. 11: B. A. Hinsdale, *President Garfield and Education.* Memorial Address at Hiram, Ohio, College (New York, Houghton Mifflin, n.d.).

Page 18   5: H. W. Dickinson, *James Watt and the Steam Engine* (Oxford, Clarendon Press, 1927), p. 222. 6: *Ibid.*, frontispiece. 7: J. F. Racknitz, *Uber den schachspieler des herrn von Kempelen und dessen nachbildung* (Dresden, 1798), folding plate. 9: John Farey, *Treatise on the Steam Engine* (London, Longman, Rees, Orme, Brown, and Green, 1827), plate #XI, p. 444. 10: J. B. L. Foucault, *Recueil des travaux scientifiques* (Paris, J. Bertrand, 1878). 11: Samuel Butler, *Erewhon and Erewhon Revisited* (New York, Modern Library, 1955), p. 237. 12: Culver Pictures, Inc.

Page 19   5b: B. V. Bowden, *Faster Than Thought* (New York, Pitman, 1953), p. 368. 9: Culver Pictures, Inc. 12: *Popular Science Monthly* 11:745-747, (Oct. 1877).

Page 22   5: The Carpenter Center for the Visual Arts, Photography Collection, Harvard University, Cambridge, Mass. 12: *Rough Count Eleventh Census of the United States* (Washington, Government Printing Office, 1890), cover.

Page 23   5: "The Eleventh Census," *Leslie's Illustrated Newspaper* 69:181-182 (Oct. 12, 1889). 5a: State Library, Commonwealth of Massachusetts, Boston. 6: Columbiana Collection, Columbia University, New York, N.Y. 6b: Letter from Herman Hollerith to his wife dated Aug. 24, 1895. 7: Report of the Commission appointed by the Honorable Superintendent of Census, "Different Methods of Tabulating Census Data Submitted December 2, 1889" (Washington, Judd Detweiler, n. d.). cover. 11: U.S. Patent #395, 781 (Jan. 8, 1889).

Page 24   5: *Scientific American* 63:cover (Aug. 30, 1890). 7: "The Eleventh Census of the United States, Schedule #1, Population and Social Statistics" (Washington, U.S. Government, n.d.). 11: T. C. Martin, "Counting a Nation by Electricity," *Electrical Engineer* 12:521-530 (Nov. 11, 1891).

Page 25   5: Library of Congress, Washington, D.C. 13: Martin, "Counting a Nation by Electricity."

Page 26   5: Library of Congress. 7: Letter from C. A. Springer to T. J. Watson, Jr., Dec. 21, 1965, IBM Archives.

Page 27   2: John H. Blodgett, *Herman Hollerith, Data Processing Pioneer* (Philadelphia, Drexel Institute of Technology, 1968), p. 69. 5: Library of Congress. 8: Keystone Press Agency, Inc., New York. 11: Book on Hollerith System in Russia, 1894, IBM Archives. 12: H. Zemanek, Vienna, Austria. 16: "Census in Rural Russia," *New York Times* 46:5 (July 12, 1897).

Page 28   1: *Anthropometric Laboratory Arranged by Francis Galton, F.R.S., for the Determination of Height, Weight, Span, Breathing Power, Strength of Pull and Squeeze, Quickness of Blow, Hearing, Seeing, Colour Sense and Other Personal Data* (London, William H. Clowes and Sons, 1884). 5, 7: Karl Pearson, *The Life, Letters and Labours of Francis Galton* (Cambridge, Cambridge University Press, 1930), vol. II. Plates.

Page 29   2, 5, 7, 8: Karl Pearson, *The Life, Letters and Labours of Francis Galton.* 11: Armed Forces Institute of Pathology, Washington, D.C.

Page 30   5, 9, 12: Brown Brothers. 6, 14: Alphonse Bertillon, *Identification of Criminals* (Chicago, American Bertillon Prison Bureau, 1889). 7: Finger Print Kit, The Galton Laboratory. 11: Francis Galton, *Finger Prints* (London, Macmillan, 1892). frontispiece.

Page 31   Gore material courtesy The Prudential Insurance Company of America, Newark, N.J.

Page 32   Marquand material courtesy Princeton University Library, Princeton, N.J. 13: Allan Marquand, "A New Logical Machine," *Proceedings of the American Academy of Arts and Sciences,* N.S. 13:303 (1886). 15: Lewis Carroll, *Mathematical Recreations of Lewis Carroll, Symbolic Logic and the Game of Logic* (New York, Dover Publishers Inc. and Berkeley Enterprises, 1958), I 118.

Page 33   5: Helen Ellis, Westport Point, Mass.; Princeton University Library. 7, 11, 15: Annibale Pastore, *Logica Formale: dedotta della considerazione di modelli meccanici* (Turin, Bocca, 1906). 9: Harvard University Archives, Widener Library, Cambridge, Mass.

Page 34   5: William K. David, *David's New Lightning Calculator* (Buffalo, N.Y., W. K. David, 1881), cover. 6: *Some Experiences with Figures* (Detroit, Buroughs Adding Machine Co., 1907), p. 2. 7: Solomon Stone, *Solomon Stone's Marvelous Mental Calculator* (New York Popular Publishing Co., 189?), cover. 9: Warshaw Collection, Smithsonian Institution. 13: *Some Experiences with Figures* (Detroit, Burroughs Adding Machine Co., 1907). 16: Dorr E. Felt. "Mechanical Arithmetic," *Scientific American* 69:310-311 (Nov. 11, 1893). Oliver Wendell Holmes, *The Autocrat of the Breakfast Table* (Boston and New York, Houghton Mifflin, 1891), p. 9. *Some Experiences with Figures,* inside front cover.

Page 35 | Léon Bollée material courtesy Elizabeth Binford, Powhatan, Va. **6**: *Album of the Paris Exhibition,* 1889 (Richmond, Va., Allen and Ginter, 1889), cover.

Page 37 | **1**: "The Millionaire" booklet (New York, W. A. Heim, n.d.). **5**: U.S. Patent #538, 710 (May 7, 1895). **7, 15**: *New York Herald,* 1878. **11**: U.S. Patent #207, 918 (Sept. 10, 1878).

Page 38 | **8**: Culver Pictures, Inc. **11**: Warshaw Collection, Smithsonian Institution, The Victor Comptometer Corp. **15**: J. A. V. Turck, *Origin of Modern Calculating Machines* (Chicago, Western Society of Engineers, 1921), p. 55.

Page 39 | **1**: Turck, *Origin of Modern Calculating Machines.* **5**: Brown Brothers. **6, 11**: Burroughs Adding Machine Co., Detroit, Mich. **9**: U.S. Patent #388,116 (Aug. 21, 1888).

Page 40 | Michelson material courtesy Ryerson Laboratory, Univ. of Chicago. **5**: A. Galle, *Mathematische Instrumente* (Leipzig, B. G. Teubner, 1912). **15**: Robert A. Millikan, *Commemorating Michelson Laboratory Dedication* (China Lake, Calif., United States Ordnance Test Station, May 1949), p. 17.

Page 41 | Michelson material courtesy Ryerson Laboratory, Univ. of Chicago. **5**: *Evening Star,* Washington, D.C. (Mar. 11, 1901), p. 59.

Page 42 | *Dawn of the Century* (New York, E. T. Paull Music Co., 1899).

Page 44 | **1**: Henry Adams, *The Education of Henry Adams: An Autobiography* (New York, Heritage Press, 1942), p. 434. **5**: L. Baschet, *Plan pratique de l'exposition universelle de 1900 contenant tous les palais et pavillons, souvenirs de l'exposition:* **8**: Culver Pictures, Inc. **9**: *L'Exposition de Paris* (Paris, Editions Montchrestien, n.d.), photo of dynamos et moteurs (selections francaises), le palais de l'electrique au Champ de Mars. **11**: Yves Delaporte, *La Cathédrale de Chartres* (Paris, Publications Filmées d'Art et d'Histoire, Edition de Chene, 1943). **15**: Adams, *Education of Henry Adams,* p. 361 **16**: Ambrose Bierce, *The Collected Works of Ambrose Bierce,* vol. II: *Can Such Things Be* (New York, Neale Publishing Co., 1910).

Page 45 | Binet materials courtesy les demoiselles Binet, Paris. **1**: Alfred Binet, *L'Etude experimentale de l'intelligence* (Paris, Schleicher Frères et Cie., 1903). **5**: Alfred Binet, *Les idées modernes sur les enfants* (Paris, Flammarion, 1909), cover.

Page 46 | **5**: Metropolitan Life Insurance Co., New York.

Page 47 | **5**: Brown Brothers.

Page 48 | **5, 6, 9**: Eaton Corp., Cleveland, O. **13**: "The Mechanical Accountant," *Engineering* 74:840-841, (Dec. 26, 1902).

Page 50 | **5a**: Elliott Frog and Switch Co. **5b**: Chart, Willard C. Brinton, *Graphic Methods for Presenting Facts* (London, Works Management Library Quarterly, 1914), p. 52. **7**: Four photos, The Pennsylvania Steel Company's *Frog Shop Digest* (Mar. 1911); Bethlehem Steel Co., Bethlehem, Pa. **9**: Morrell W. Gaines, "Tabulating-Machine Cost-Accounting for Factories of Diversified Product," *The Engineering Magazine* 30:364-373 (Dec. 1905). **11**: Gershom Smith, "Distribution of Indirect Costs by the Machine Hour Method," *The Engineering Magazine* 37:384 (Jun. 1909). **12**: New York Public Library Picture Collection; Bethlehem Steel Co.

Page 51 | **5**: Brown Brothers.

Page 52 | **7, 12**: Slide rules, courtesy Graduate School of Business Administration, Harvard University, Cambridge, Mass. **9**: Brown Brothers. **13**: Frederick W. Taylor, *The Art and Science of Shoveling,* U.S. Congress. House. Special Committee to investigate the Taylor and other systems of shop management, 1912.

Page 53 | **1**: L. Frank Baum, *The Road to Oz,* (Chicago, Illinois, Reilly and Lee, 1909), p. 171. **5**: Brown Brothers. **6**: L. F. Baum, *Tik-Tok of Oz,* (Chicago, Ill., Reilly and Lee Co., Rep. 1936). cover. **7**: L. F. Baum, *The Road to Oz,* (Chicago, Ill., Reilly and Lee Co., 1909), p. 171. **9**: L. F. Baum, *The Road to Oz,* p. 154.

Page 54 | **5**: Lowell Observatory, Flagstaff, Ariz.

Page 55 | **5**: W. Jordan, *Opus Palatinum Sinus-und Cosinus-Tafeln von 10 zu 10* (Hannover, Hahnische Buchhandlung, 1897). cover. **7**: Museum of History of Science, Oxford, Eng. **8**: A. Crelle, *Doctor A. Crelles Calculating Tables* (Berlin, G. Riemer, 1902). cover. **13**: L. J. Comrie, "The Application of Calculating Machines to Astronomical Computing," *Popular Astronomy* 33:2-8 (Apr. 1935).

Page 56 | **5, 7**: Lowell Observatory. **8**: A. L. Lowell, *The Biography of Percival Lowell* (New York, Macmillan, 1935). **16**: R. L. Putnam, "Searching Out Pluto, Lowell's Transneptunian Planet X," *Scientific Monthly* 34:5-21 (Jan. 1932).

Page 57 | **5**: Leverin and Co., Stockholm, Sweden. The National and University Library in Oslo, Norway. **7**: Vilhelm Bjerknes, *Dynamic Meteorology and Hydrography* (Washington, D.C., Carnegie Institute, 1905). **9**: *Meteorologische Zeitung* 21: cover (1904). **10**: Leverin and Co., Stockholm. **15**: V. Bjerknes, Inaugural Lecture at Installation as Professor in the new chair of Geophysics, University of Leipzig, Jan. 8, 1913, *Monthly Weather Review* 42:11-14 (1914).

Page 58 | 164th Depot Brigade, Camp Funston, Kan.

Page 60 | **5**: "The Modern Torpedo," *Scientific American* 90:196 (Mar. 5, 1904). **9, 10, 13**: W. G. Fitz-Gerald, "The Torpedo in Modern Warfare," *Harper's Weekly* 49:602-605 (Apr. 29, 1905). **11**: "The Gyroscope as a Compass," *Scientific American* 96:294-295 (Apr. 6, 1907). **12**: Preston R. Bassett, Ridgefield, Conn.

Page 61 | **7**: *Sperryscope* 7:2 (Oct. 1933). **11**: "The Regnard Aeroplane," *Scientific American* 103:228-229 (Oct. 8, 1910).

Page 62 | **5**: *Sperryscope* 7:#7, (Apr. 1935). **7**: John W. R. Taylor, *A Picture History of Flight* (New York, Pitman, 1956). **9**: J. J. Ide, "The Sperry Gyroscopic Stabilizer," *Scientific American* 111:96 (Aug. 8, 1914). **11**: Elmer Sperry's Gyrostabilizer, National Air and Space Museum, Smithsonian Institution, Washington, D.C.

Page 63 | **3, 7**: Frederick W. Lanchester, *Aircraft in Warfare* (London, Constable, 1916; New York, D. Appleton, 1917). **5**: Bombsight, United States Naval Academy Museum, Annapolis, Md.

Page 64 | **5, 10**: Russian Odhner, National Museum of History and Technology, Smithsonian Institution, Washington, D.C. **6**: Odhner History: *An Illustrated Chronicle of "A Machine to Count On"* (Goteborg, Sweden, Aktiebolaget Original Odhner, 1951), p. 54. **9**: U.S. Patent #209, 416 (Oct. 29, 1878). **12**: *Odhner History,* p. 29.

Page 65 | **5, 8**: Ford Archives, Dearborn, Mich. **12**: Culver Pictures, Inc.

Page 66 | **7**: "Torres and His Remarkable Automatic Devices," *Scientific American* 113:296 (Nov. 6, 1915). **10**: Leonardo Torres y Quevedo, "Ensayos sobre

1	2	3	4
5	6	7	8
9	10	11	12
13	14	15	16

Automaticos," *Real Academia de Ciencias Exactas, Fiscias y Naturales, Revista* 12:391 (1913).

Page 67   2: "Torres and His Remarkable Automatic Devices." 5: Studio Constantin, Paris, France.

Page 68   3: Leopoldo Rodriquez Alcalde, *Leonardo Torres Quevedo y la cibernetica* (Madrid, 1966). 5, 6, 9: Leonardo Torres-Quevedo, Madrid. 7: No-end Axle Machine, Escuela de Ingenieros de Caminos, Madrid.

Page 69   1: R. A. Harris, "The Coast and Geodetic Survey Tide Predicting Machine," *Scientific American* 110:485 (Jun. 13, 1914). 5, 11: NOAA, National Ocean Survey, Department of Commerce, Washington, D.C. 7: Underwood and Underwood, New York. 8: C. H. Claudy, "A Great Brass Brain," *Scientific American* 110:197-198 (Mar. 7, 1914). 12: H. Rauschelbach, "Die Deutsche Gezeitenrechenmaschine," *Zeitschrift Fur Instrumentenkunde* 44:285-303 (Jul. 1924).

Page 70   Pearson material courtesy E. S. Pearson, London. 1: History of the Biometric and Galton Laboratories 1920 Statement in connection with the opening of the new building given by Sir Herbert H. Bartlett to house the Department of Applied Statistics, University College London, Eng. 8: *Biometrika: A Journal for the Statistical Study of Biological Problems* 1: cover (1901). 12: W. P. Elderton, "Tables for Testing the Goodness of Fit of Theory to Observation," *Biometrika* 1: 155-163 (Jan. 1902).

Page 71   1: K. Pearson, "On the Criterion that a Given System of Deviations from the Probable in the Case of a Correlation System of Variables is such that it can reasonably be supposed to have arisen from Random Sampling," *Philosophical Magazine* 50:157-175, series 6 (1925). 5, 13: Eleanor Pairman, *Tables of the Digamma and Trigamma Function,* vol. 1 of *Tracts for Computers,* edited by Karl Pearson, F. R., (London, Cambridge University Press, 1919). cover. 8: E. S. Pearson, London. 11: Radio Times Hulton Picture Library, London. 16: K. Pearson, ed., *Tables for Statisticians and Biometricians* (Cambridge, Cambridge University Press, 1930, 1931).

Page 72   8: New York Life Insurance Co., New York. 9, 11: Bureau of the Census, Washington, D.C.

Page 73   5: Library of Congress

Page 74   5: National Archives, Washington, D.C.

Page 75   5: National Archives

Page 76   5: Ben D. Wood, New York, N.Y. 8: Geraldine Joncich, *The Sane Positivist: A Biography of E. L. Thorndike* (Middletown, Conn., Wesleyan University Press, 1968), facing p. 22. 9: Pictorial Competition Test—1, by William Healy, C. H. Stoelting and Co., Chicago, Ill. 10: National Archives. 12: *Examiner's Guide for Psychological Examining in the Army* (Washington, D.C., Government Printing Office, 1918), cover.

Page 77   5: Archives of the History of American Psychology, Akron, O. 7: *The Medical Department of the United States Army in the World War,* vol. 15, *Statistics,* Part I, "Army Anthropology" (Washington, D.C., War Department, 1921), p. 61. 9: "The Measurement and Utilization of Brain Power in the Army, II," *Science* 49:251-259 (Mar. 14, 1919). 11: Culver Pictures, Inc. 13: Letter from H. H. Goddard to Lewis M. Terman, Nov. 25, 1918.

Page 78   Aberdeen material courtesy Mrs. Norbert Wiener, South Tamworth, N.H. 12, 13, 16: Norbert Wiener, *Ex-Prodigy: My Childhood and Youth* (New York, Simon and Schuster, 1953), pp. 255, 257, 258.

Page 79   5: University of Chicago Library, Chicago. 6: Library of Congress. 7: W. C. Nelson, ed., *Selected Topics on Ballistics: The Cranz Centenary Colloquium, University of Freiburg* (Oxford, Pergamon Press, 1959), p. 11. 9: Forest R. Moulton, *New Methods in Exterior Ballistics* (Chicago, University of Chicago Press, 1926), cover. 11: Ballistics Research Laboratory, Aberdeen Proving Ground, Md. 13: "ENIAC," *Ordnance Magazine* 54:185 (Sept.-Oct. 1969).

Page 80   5: Mrs. Michael Traylen, Devon, Eng. 7, 15: Lewis F. Richardson, *Weather Prediction by Numerical Process* (Cambridge, Cambridge University Press, 1922), p. 220. 9: S. A. Richardson, Riverside, Conn. 12, 16: Richardson, *Weather Prediction,* p. 136.

Page 81   2: Richardson, *Weather Prediction,* preface. 5: Lewis F. Richardson, *Mathematical Psychology of War,* (Castlehead, Eng., L. F. Richardson, 1919). 6, 10: Mrs. Michael Traylen. 7, 8: S. A. Richardson. 16: Tribute by Dorothy Richardson published in *Paisley and Renfrewshire Gazette* (Oct. 3, 1953).

Page 84   1: Letter from Vannevar Bush to I. J. Seligsohn, Feb. 11, 1970. 5: Profile Tracer, Vannevar Bush, Belmont, Mass.

Page 85   2: *New York Evening Post,* vol. 126, no. 291 (Oct. 27, 1927). 5, 8: Photo and Watt Hour Meter, National Museum of History and Technology, Smithsonian Institution, Washington, D.C.

Page 86   1: L. J. Comrie, "The Application of Calculating Machines to Astronomical Computing," *Popular Astronomy* XXXIII: 2-8 (Apr. 1935). 5, 7, 10: Mrs. B. Atkinson, Stroud, Eng. 8: *The Nautical Almanac and Astronomical Ephemeris for the Year 1931* (London, H. M. Nautical Almanac Office, 1930), cover. 9: L. J. Comrie, *The Hollerith and Powers Tabulating Machines* (London, printed for private circulation, 1933), cover. 11a: L. J. Comrie, "Inverse Interpolation and Scientific Applications of the National Accounting Machine," *Journal of the Royal Statistical Society* 3: cover (1926). 11b: L. J. Comrie, "German Calculating Machine Enterprise," (London, Office Machinery Users Association, 1929). 12: L. J. Comrie, *Computing by Calculating Machine* (London, Gee and Co., 1927), cover. 14: L. J. Comrie, "Computing by Calculating Machine," *Accountants Journal* XLV:42-51 (May 1927).

Page 87   1: E. W. Brown, *Tables of the Motion of the Moon,* 3 vols. (New Haven, Yale University Press, 1920). 3: L. J. Comrie, *Modern Babbage Machines* (London, Office Machinery Users Association Ltd. Bulletin, n. d.). 5: A. J. Thompson, *Logarithmetica Brittanica,* vol. I (Cambridge, Cambridge University Press, 1957), frontispiece. 7: National Cash Register Co., Dayton, O. 11: Burroughs Calculator, National Museum of History and Technology, Smithsonian Institution.

Page 88   2: Paul de Kruif, *Hunger Fighters* (Harcourt, Brace and World, N.Y., 1928), p. 222. 5: University of Iowa Library, Iowa City, Ia. 8: Henry A. Wallace, *Correlation and Machine Calculation* (Ames, Ia., Iowa State University Press, 1925), cover. 10, 11: Pioneer Hi-Bred International, Des Moines, Ia.

168

Page 89   5: *Wallaces' Farmer* 6: cover (Feb. 4, 1921). 7: Stuart C. Dodd, Seattle, Wash. 8: *Industrial Psychology* 1: cover (Jan. 1926). 10: ''Machine Does Year's Work in Day,'' *Baltimore Daily Post*, no. 716 (Mar. 9, 1925). 14: Clark L. Hull, ''An Automatic Correlation Calculating Machine,'' *Journal of the American Statistical Association* 20:522-531 (Dec. 1925).

Page 90   12, 15: Thomas J. Watson, Sr., *Men, Minutes, and Money* (New York, IBM, 1934).

Page 91   3: Watson, *Men, Minutes, and Money*, p. 196.

Page 93   11: *System: The Magazine of Business* (Amer. ed.) 50: cover (Aug. 1926). 15: Thomas and Marva Belden, *The Lengthening Shadow* (Boston, Little Brown & Co., 1962), p. 169.

Page 94   1: E. L. Thorndike, ''The Nature, Purposes, and General Methods of Measurement,'' *17th Yearbook of the National Society for the Study of Education* (Chicago, University of Chicago Press, 1918), p. 16.

Page 95   5: ''Super Computing Machines Shown,'' *New York World* (Mar. 1, 1920). 6, 10: W. S. Learned and Ben D. Wood, *The Student and his Knowledge* (New York, Carnegie Foundation for the Advancement of Teaching, Bulletin 29, 1938), cover, p. 135. 7: Educational Testing Service, Princeton, N.J.

Page 96   8, 12: M. Ilin (pseud.), *The Story of the Five-Year Plan: New Russia's Primer* (Boston, Houghton Mifflin, 1931).

Page 97   5, 7: The Ford Archives, Dearborn, Mich. 9: ''50 Russians to Learn Tractor,'' *Ford News* 6:1 (Mar. 22, 1926), courtesy of The Ford Archives. 10, 11: G. T. Grinko, *The Five-Year Plan of the Soviet Union* (New York, International Publishers, 1930).

Page 98   1, 11: Nicholas Minorsky, ''Directional Stability of Automatically Steered Bodies,'' *Journal of the American Society of Naval Engineers* 34:280-309 (May 1922). 6: ''Our Latest Dreadnought, the 'New Mexico,' '' *Scientific American* 115:383 (Oct. 28, 1916). 8: ''Metal Mike,'' *Sperryscope* 7:5 (Jul. 1933). 9: *Sperryscope*, vol. 7 (Jan. 1933).

Page 99   2: Walter B. Cannon, *The Wisdom of the Body* (New York, Norton, 1932; rev. and enlarged, 1939). 3: Norbert Wiener, *Cybernetics: Or Control and Communication in the Animal and the Machine* (Cambridge, Mass., M.I.T. Press, 1948). 5, 12: Harvard Medical Library, Francis A. Countway Library of Medicine, Boston, Mass. 10, 11, 15: W. B. Cannon, ''Organization for Physiological Homeostasis,'' *Physiological Reviews* 9:399-431 (1929).

Page 100   Radio Times Hulton Picture Library, London.

Page 102   5, 9: ''RUR, A Dramatic Indictment of Civilization,'' *Current Opinion* 74:61-74 (Jan. 1923), courtesy *Time* magazine. 7, 11: Universum Film Aktiengesellschaft, Germany, 1927; Academy of Motion Picture Arts and Sciences, Hollywood, Cal. 12: E. M. Forster, *The Eternal Moment* (New York, Harcourt, Brace and World, 1928), cover.

Page 103   5: Film Strip, Universum Film Aktiengesellschaft. 7: Universum Film Aktiengesellschaft; Academy of Motion Picture Arts and Sciences.

Page 104   5a: Aldous Huxley, *Brave New World* (New York, Doubleday Doran and Co., 1932) p. 1. 5b: Modern Times 1936, Columbia Pictures; Museum of Modern Art/Film Stills Archives, New York, N.Y. 7: ''A Challenge to Technocracy,'' *The New York Times Magazine* 82:1, sec. 6 (Jan. 8, 1933). 10: *A Nous la Liberté*, Theatre Collection, The New York Public Library at Lincoln Center, New York, N.Y. 13: *The New York Herald Tribune*, (Feb. 6, 1936). 15: H. Scott, *Energy Survey of North America* (Rushland, Pa., Technocracy, Inc., 1933). Archibald MacLeish, ''Machines and the Future,'' *Nation* 136:140-142 (Feb. 8, 1933).

Page 105   5, 7: United Press International, New York, N.Y. 9: Alfred Chapuis, *Automata: A Historical and Technological Study* (London, V. T. Batsford Ltd., 1958), p. 384; originally published by Editions du Griffon, Neuchatel, Switzerland. 10: Westinghouse Electric, New York, N.Y. 13: ''The Electric Tail-Wagger,'' *Newsweek* 15:42 (May 13, 1940).

Page 106   5: ''The Machine That Bosses Other Machines,'' *Literary Digest* 86:20 (Sep. 9, 1933); courtesy *Time* Magazine. 8: Pennzoil, Houston, Tex. 12: Loaned by R.E.S.I.S.T.O.R.S. from the Collection of Claude A. R. Kagan, Pennington, N.J.

Page 107   3: H. L. Hazen, ''The Theory of Servo-Mechanisms,'' *Journal of the Franklin Institute* 218:279-331 (Sept. 1934). 5: H. S. Black, Summit, N.J. 6: *Ten Men and the Telephone* (Murray Hill, N.J., Bell Telephone Laboratories, 1961). p. 22. 7: Taylor Automatic Controller, Sybron Corp., Rochester, N.Y. 9: Black Amplifier, American Telephone and Telegraph Co., New York, N.Y.

Page 108   1: Campaign Speech, Franklin D. Roosevelt, Detroit, Mich. Oct. 2, 1932. 5, 7: United Press International, New York, N.Y. 9: Social Security Administration, Baltimore, Md., M. E. Poole. 10: Brown Brothers.

Page 109   5: ''Biggest Bookkeeping Job Begins,'' *Sunday News* 16, no. 40, (Jan. 10, 1937), p. 60. 9, 10: *Time-Life-Fortune 540* (New York, Time Inc., n.d.); photographs of ''Sorter'' and ''Choosey'' reprinted by permission from Time, The Weekly Newsmagazine; ©Time Inc. 12: Social Security Administration, Baltimore, Md. 15: Discussion meeting with H. J. McDonald, Aug. 1969.

Page 110   5, 8b: Peabody Museum, Harvard University, Cambridge, Mass. 7: E. A. Hooton, *The American Criminal* (Cambridge, Mass., Harvard University Press, 1939), cover. 8a, 10: A. Griffin, Winchester, Mass. 13: E. A. Hooton, ''Calipers and Criminals,'' *Harvard Alumni Bulletin* 33:344-351 (Dec. 11, 1930), 15: E. A. Hooton, *Why Men Behave Like Apes and Vice Versa*, (Princeton, N.J., Princeton University Press, 1940); E. A. Hooton, *Young Man, You Are Normal*, (N.Y. Putnam, 1945).

Page 111   5: American Institute of Public Opinion, Princeton, N.J. 8: Theodore H. Brown, *Use of Statistical Techniques in Certain Problems of Market Research* (Cambridge, Mass., Harvard University Bureau of Business Research, 1935). 9: Wide World Photos, Inc., New York, N.Y. 10: ''Landon Over Roosevelt Final Returns,'' *Literary Digest* 122:5-6 (Oct. 31, 1936), courtesy *Time* Magazine. 11 ''What Went Wrong at the Polls,'' *Literary Digest* 122:7-8 (Nov. 14, 1936), courtesy *Time* Magazine.

Page 112   5: Office of Business Economics, Department of Commerce, Washington, D.C. 8, 12: Wassily W. Leontief, Cambridge, Mass. 13: Discussion meeting with I. Bernard Cohen, Owen Gingerich, and Wassily Leontief, Apr. 1969. 16: W. Leontief, ''Interrelation of Prices, Output, Savings and Investment,'' *Review of Economic Statistics* 19:109-132 (Aug. 1937).

Page 113   5: Plate from machine, John V. Wilbur, M.I.T., Cambridge, Mass. 8: John V. Wilbur, ''Mechanical Solution of Simultaneous Equations,'' *Journal of the*

1	2	3	4
5	6	7	8
9	10	11	12
13	14	15	16

170

Franklin Institute 222:715-724 (Dec. 1936). **14**: Discussion meeting with I. Bernard Cohen, Owen Gingerich, and Wassily Leontief, Apr. 1969.

Page 114   **1**: Wallace J. Eckert, "Early Computers," *IBM Research News* (May 1963), p. 7. **5**: Wallace J. Eckert, James Eckert, Washington, D.C. **13**: Howard K. Janis, "Dr. Wallace J. Eckert Retires," *IBM News National Edition* 4:5 (Jul. 25, 1967).

Page 115   Eckert material courtesy Mrs. Wallace J. Eckert, Leonia, N.J. **13**: Janis, "Dr. Wallace J. Eckert Retires."

Page 116   **5**: National Museum of History and Technology, Smithsonian Institution. **13**: Vannevar Bush, *Pieces of the Action* (New York, William Morrow, 1970, p. 161.

Page 117   **5**: Edwin L. Harder, Pittsburgh, Pa. **7**: United Press International. **10, 12**: Gordon S. Brown, M.I.T.

Page 118   **5**: E. L. Harder. **7**: Svein Rosseland, "Mechanische Integration von Differentialgleichungen," *Die Naturwissenschaften* 27:729-735,(Nov. 3, 1939). **8**: J. P. Madden Inc., Bethlehem, Pa. **9**: C. L. Beuken, "Warmeverluste bei Periodish Betriebenen Elektrischen Ofen," Inaugural dissertation, University of Freiburg, 1936. **10**: L. Beuken, Maastricht, Netherlands.

Page 119   **2**: D. R. Hartree, "Approximate Wave Functions and Atomic Field for Mercury," *Physical Review* 46:738-743 (Oct. 15, 1934). **7**: Science Museum, London; courtesy Guttenberg Ltd. **9**: Meccano Differential Analyzer, Science Museum, London. **11, 15**: Arthur Porter, "The Construction of a Model Mechanical Device for the Solution of Differential Equations with Applications to the Determination of Atomic Wave Functions," Master's thesis, Victoria University of Manchester, Sep. 1934; D. R. Hartree, "The Mechanical Integration of Differential Equations," *The Mathematical Gazette* 22:349 (Oct. 1938).

Page 120   Zuse material courtesy Konrad Zuse, Hünfeld, Germany.

Page 121   **1**: George R. Stibitz, "The Relay Computers at Bell Labs," *Datamation* 13:35-44 (May 1967). **4, 8**: Bertrand Russell and A. N. Whitehead, *Principia Mathematica* (Cambridge, Cambridge University Press, 1910). **5, 6**: Bell Telephone Laboratories, Murray Hill, N.J. **9, 13**: C. E. Shannon, "A Symbolic Analysis of Relay and Switching Circuits," *Transactions of the AIEE* 57:713-723 (Dec. 1938). **10**: George Stibitz, Hanover, N.H. **16**: C. E. Shannon, "A Symbolic Analysis of Relay and Switching Circuits," Master's thesis, M.I.T., 1937.

Page 122   **1**: Memorandum re: Computing Mechanisms for Harvard University by Dean H. M. Westergaard, Howard Aiken, and J. W. Bryce, Jan. 18, 1939. **5, 7, 9**: Cruft Laboratory, Harvard University, Cambridge, Mass.

Page 123   **5, 6**: Cruft Laboratory. **9**: "Harvard's Robot Super-Brain," The *American Weekly,* Oct. 15, 1944 **15**.: L. J. Comrie, "Babbage's Dream Comes True," *Nature* 158:567 (Oct. 26, 1946).

Page 124   **1**: A. M. Turing, "On Computable Numbers with an Application to the Entscheidungsproblem," *Proceedings London Mathematical Society* 42:230-265 (July 1937). **5, 13**: Letter from A. M. Turing to his mother, Ethel T. Turing, May 29, 1936; Kings College, Cambridge. **8**: Bassano and Vandyk Studios, London. **11**: H. Wang, "Games, Logic and Computers," *Scientific American* 213:98-106 (Nov. 1965). **15**: J. G. Kemeny, "Man Viewed as A Machine," *Scientific American* 192:58-67 (Apr. 1955).

Page 125   **1**: Turing, "On Computable Numbers." **3**: M. H. A. Newman "A. M. Turing 1912-1954" *Royal Society London Biographical Memoirs* 1:252-263 (1955). **5, 9**: Kings College, Cambridge. **6**: Turing, "On Computable Numbers," p. 230. **8, 12**: Emil Post, "Finite Combinatory Processes Formulation," *Journal of Symbolic Logic* 1:103-110 (Sep. 1936).

Page 126   National Archives.

Page 128   **5**: C. A. Lovell, McLean, Va. **7**: Operational amplifier, D.C. Bomberger, Bell Telephone Laboratories, Whippany, N.J. **9**: D. B. Parkinson, Cleveland, O.

Page 129   **6**: Telegram from Norbert Wiener to H. H. Goldstine. **7**: Norbert Wiener, *The Extrapolation, Interpolation, and Smoothing of Stationary Time Series with Engineering Applications* (Cambridge, Mass., M.I.T., 1942), cover; researched on behalf of the National Defense Research Council. **11**: Julian Bigelow, Princeton University, Princeton, N.J. **14**: Letter to H. H. Goldstine from Norbert Wiener dated Dec. 28, 1944. **15, 16**: Wiener, *Extrapolation, Interpolation and Smoothing.*

Page 130   **5**: Simulation Products Division, The Singer Co., Binghamton, N.Y. **7, 11**: Gordon S. Brown, M.I.T.

Page 131   Project Pigeon materials, B. F. Skinner, Cambridge, Mass.

Page 132   **5**: United Press International.

Page 133   ENIAC Materials, Historical Services Division, Department of the Army, Washington, D.C.; **16**: "A Time to Think and Freedom to Explore." *IBM News Magazine* 1:10-15 (Jul. 8, 1969).

Page 134   **1**: "Report on an electronic diff*analyzer" submitted to the Ballistic Research Laboratory, Aberdeen Proving Grounds, by the Moore School of Electrical Engineering, University of Pennsylvania, Apr. 2, 1943. **5**: *Baltimore Evening Sun* Photo Collection, Baltimore, Md. **8**: D. C. Bomberger, Bell Telephone Laboratories. **9**: H. H. Goldstine, Princeton, N.J.

Page 135   **5**: B. V. Bowden, *Faster Than Thought* (London, Sir Isaac Pitman and Sons Ltd., 1953), frontispiece. **6**: *Philadelphia Evening Bulletin* Photo Collection, Philadelphia, Pa. **7**: Wide World Photos, Inc. **9**: H. H. Goldstine.

Page 136   **5**: Alan W. Richards, Princeton, N.J., 1944. **13**: Proposal on IAS machine (John von Neumann).

Page 137   **5**: W. S. McCulloch and W. Pitts, "A Logical Calculus of the Ideas Imminent in Nervous Activity," *Bulletin of Mathematical Biophysics* 5:115-133 (Dec. 1943). **6**: H. H. Goldstine. **8**: Selectron tube, National Museum of History and Technology, Smithsonian Institution. **9**: John von Neumann, *Theory and Techniques for Design of Electronic Digital Computers,* vol. I, First Draft of a report on the EDVAC (Philadelphia, University of Pennsylvania, Jun. 1945), cover. **16**: A. W. Burks, H. H. Goldstine, and John von Neumann, *Preliminary discussion of the logical design of an electronic computing instrument* (Princeton, N.J. Institute for Advanced Study, 1945), vol. I, part I, p. 5.

Page 138   Von Neumann—Goldstine material courtesy H. H. Goldstine.

Page 139   **5**: Joseph Smagorinsky, Princeton University. **6**: Letter from L. F. Richardson to Jule Charney, Jun. 19, 1952; courtesy S. A. Richardson, Riverside, Conn. **8, 12**: Jule Charney, "Numerical Integration of the Barotropic Vorticity Equation," *Tellus* 2:237-254 (1950).

Page **140** 5: George Stibitz, Hanover, N.H. 6: Bell Telephone Laboratories, Murray Hill, N.J.

Page **141** 5, 9: Photos courtesy Bell Telephone Laboratories. 7: Teletype Tape and Reader, A.T. & T., New York, N.Y.

Page **142** 5: Marina Whitman, Council of Economic Advisors, Washington, D.C.; E. W. Paxson, Rand Corp., Santa Monica, Cal. 7, 11: Marina Whitman. 8, 15: Oskar Morgenstern and John von Neumann, *Theory of Games and Economic Behavior* (Princeton, Princeton University Press, 1944).

Page **143** 5: Philip M. Morris, *Methods of Operations Research* (Cambridge, Mass., M.I.T. Press, 1951), p. 126. 7: T. C. Koopmans, S. Reiters, *The Tanker Freight Rates and Tank Ship Building* (Haarlen, Publication #27, Netherlands Economic Institute, 1939), p. 246.

Page **144** 1, 4: Francis Bello, "The Information Theory," *Fortune* 23:136-158 (Dec. 1953). 5: C. E. Shannon, CBS Television Network, New York, N.Y. 6: *Bell System Technical Journal* 27: cover (Jul. 1948). 7: C. E. Shannon, "Communication Theory of Secrecy Systems," *Bell System Technical Journal* 28:656-715 (Oct. 1949).

Page **145** 3: W. F. Friedman, *The Index of Coincidence and its Application in Cryptography* (Geneva, Ill., Riverbank Laboratories, 1922). 5: Gray Code Disk, Bell Telephone Laboratories, Holmdel, N.J. 7: United Press International. 11: Culver Pictures, Inc. 12: National Archives. 15: Bello, "Information Theory."

Page **146** 5: Tortoise, W. Grey Walter, Burden Neurological Institute, Bristol, Eng. 8: Studio Constantin, Paris. 12, 14: "Machines Turn Turtle," *Life Magazine* 28:147-151 (May 15, 1950).

Page **147** 5: M.I.T., Cambridge, Mass. 7: J. G. Kemeny, "Man Viewed as a Machine." 9: Norbert Wiener, *Cybernetics.*

Page **148** 5: W. Ross Ashby, "Design for a Brain," *Electronic Engineering* 20:379-383 (Dec. 1948). 7: Studio Constantin, Paris. 8: Film from Ford Motor Co., Dearborn, Mich. 9, 13: W. Ross Ashby, Glouster. 15: Norbert Wiener, "A Machine Wiser than its Maker," *Electronics* 26:368-378 (Jun. 1953).

Page **149** 5: A. Newell, "Chess Playing Programs and the Problem of Complexity," *IBM Journal of Research and Development* 2:320-335 (Oct. 1958). 6, 7: Bell Telephone Laboratories. 9, 11: Chess Board and Mouse Panel, Claude E. Shannon, Winchester, Mass. 14: C. E. Shannon, "Programming a Computer for Playing Chess," *Philosophical Magazine* (London) 41:256-275 (1950). 15: C. E. Shannon, "Presentation of A Maze-Solving Machine," (Transactions of the 8th Conference entitled *Cybernetics,* New York, Josiah Macy Jr. Foundation, 1951), p. 173.

Page **150** 1: R. R. Everett, "The Whirlwind I Computer," *Review of Electronic Digital Computers Joint AIEE-IRE Computer Conference, Philadelphia, Pennsylvania, December 10-12, 1951* (New York, AIEE, Feb. 1952). 5: MITRE Corp., Bedford, Mass.

Page **151** 5: Gordon S. Brown, M.I.T. 9: Magnetic core, Kenway, Jenney, and Hildreth, Boston, Mass. 12: Ceramic core array, Charles W. Adams, Burlington, Mass.

Page **152** 5: Columbiana Collection, Columbia University, New York, N.Y.

Page **153** 5: Shell Oil Co. Advertisement, *Saturday Evening Post* 223:2 (Dec. 16, 1950).

Page **154** 5: *Theory and Techniques for Design of Electronic Digital Computers: Lectures given at the Moore School,* 8 July, 1946 — 31 August, 1946, vol I, lectures 1-10 (Philadelphia, Pa., University of Pennsylvania, 1947), cover. 7: Marina Whitman, Washington, D.C.

Page **155** EDSAC materials, M. V. Wilkes, Cambridge University.

Page **156** 5: The Mansell Collection, London. 7: *The Illustrated London News,* (Jun. 25, 1949), p. 1; Wide World Photos, Inc.

Page **157** 5: Alan W. Richards, Princeton, N.J.

Page **158** 5: Commander Grace Hopper, Department of the Navy, Washington, D.C. 7: United Press International. 10: Loaned by the R.E.S.I.S.T.O.R.S., from the Collection of Claude A. R. Kagan; ©UNIVAC, Sperry Rand Corp.

Page **159** SEAC materials, Property Management Section, National Bureau of Standards, Washington, D.C.

172